TRAIPSING INTO EVOLUTION

TRAIPSING INTO EVOLUTION

INTELLIGENT DESIGN AND THE KITZMILLER V. DOVER DECISION

BY

DAVID K. DEWOLF,   JOHN G. WEST,

CASEY LUSKIN   AND   JONATHAN WITT

Description

This book is a critique of Judge John E. Jones III's controversial decision in *Kitzmiller et al. v. Dover Area School Board* (2005). The four named authors contributed to each of the book's chapters. It also includes three appendices. Appendix A is "Whether ID is Science: Michael Behe's Response to *Kitzmiller v. Dover*" by Michael J. Behe (1952–) of Lehigh University. Appendix B is "Selected Peer-Reviewed and Peer-Edited Publications Supporting the Theory of Intelligent Design (Annotated)." Appendix C is "Brief of Amici Curiae Biologists and Other Scientists in Support of the Defendants in Kitzmiller v. Dover Area School District."

Publisher's Note

This book is part of a series published by the Center for Science & Culture at Discovery Institute in Seattle. Previous books in that series include *Are We Spiritual Machines?: Ray Kurzweil vs. The Critics of Strong A.I.* by Jay W. Richards et. al., *Getting the Facts Straight: A Viewer's Guide to PBS's Evolution* by the Discovery Institute, and *Why Is a Fly Not a Horse?* by Giuseppe Sermonti.

Library Cataloging Data

Traipsing into Evolution: Intelligent Design and the Kitzmiller v. Dover Decision
Authors: David K. DeWolf (1949–), John G. West (1964–), Casey Luskin (1978–), and Jonathan Witt (1966–)
123 pages, 6 x 9 x 0.29 inches, 229 x 152 x 7.4 mm.
1. Evolution–Study and teaching–Law and legislation–United States.
2. Science and law–United States.
BISAC Subject Headings: LAW092000 LAW/Educational Law & Legislation. SCI027000 SCIENCE/Life Sciences/Evolution. EDU029030 EDUCATION/Teaching Methods & Materials/Science & Technology. EDU034000 EDUCATION/Educational Policy & Reform/General.
ISBN-10: 0-9638654-9-8   ISBN-13: 978-0-9638654-9-6

Publisher Information

Discovery Institute Press, Discovery Institute, 1511 Third Avenue, Suite 808, Seattle, WA 98101. Internet: http://www. discovery.org/

Published in the United States of America on acid-free paper.
First Edition, First Printing, March 2006.

CONTENTS

JUDICIAL COURAGE OR JUDICIAL OVERREACH?

traipse: *intr. v. To walk about idly or intrusively.*

—AMERICAN HERITAGE DICTIONARY[1]

"[T]he Court is confident that no other tribunal in the United States is in a better position than are we to traipse into this controversial area.... [W]e will offer our conclusion on whether ID is science not just because it is essential to our holding that an Establishment Clause violation has occurred in this case, but also in the hope that it may prevent the obvious waste of judicial and other resources which would be occasioned by a subsequent trial involving the precise question which is before us."[2]

"Those who disagree with our holding will likely mark it as the product of an activist judge. If so, they will have erred as this is manifestly not an activist Court."[3]

—JUDGE JOHN E. JONES III, *KITZMILLER ET AL. V. DOVER AREA SCHOOL BOARD*

As soon as it was issued at the end of 2005, the decision in *Kitzmiller v. Dover*—the first lawsuit to deal squarely with intelligent design (ID) in public schools—provoked sharp and conflicting responses. Defenders of Darwinian evolution predictably hailed the 139-page ruling as the careful treatise of an impartial and principled jurist, while critics of the

---

1. THE AMERICAN HERITAGE DICTIONARY 1285 (2nd college ed., Boston, Houghton Mifflin 1982).

2. *Kitzmiller et al. v. Dover Area School Board*, No. 04cv2688, 2005 WL 3465563, *26 (M. D. Pa. Dec. 20, 2005).

3. *Id.* at *51–*52.

theory quickly condemned the judge for being a results-oriented judicial activist.

Although it was hoped by many of the lawyers involved in the litigation that *Kitzmiller* would serve as a "test case" for intelligent design, in many respects *Kitzmiller* was wholly unsuited to serve that function.

First, the nation's leading ID proponents neither sought nor supported the policy adopted by the Dover school board. By requiring ID to be mentioned in classrooms, the school board rejected the approach of Discovery Institute, the best known supporter of intelligent design. While the Institute favors requiring students to learn about scientific evidence for and against neo-Darwinism, it opposes efforts to mandate the study of alternatives to Darwinian evolution such as intelligent design.[4] The Institute repeatedly urged repeal of the Dover policy well before any lawsuit was filed—to no avail. It continued to express its dissatisfaction with the policy throughout the course of the litigation.[5]

Second, it is now evident that most Dover school board members knew little if anything about intelligent design when they adopted their policy. The instigators of the policy were supporters of Biblical creationism, not intelligent design, and after the policy's adoption, board mem-

4. See *Discovery Institute's Science Education Policy,* at http://www.discovery. org/scripts/viewDB/index.php?command=view&id=3164&program=CS C%20-%20Science%20and%20Education%20Policy%20-%20School%20D istrict%20Policy (last visited Jan. 27, 2006); John G. West & Seth Cooper, *Discovery Institute Opposes Proposed PA Bill on Intelligent Design, Letter to Pennsylvania Legislature,* at http://www.discovery.org/scripts/viewDB/index. php?command=view&id=2688 (last visited Jan. 27, 2006).

5. Pennsylvania School District Considers Supplemental Textbook Supportive of Intelligent Design (dated Oct. 6, 2004), *at* http://www.discovery. org/scripts/viewDB/index.php?command=view&id=2231 (last visited Jan. 27, 2006); Discovery Calls Dover Evolution Policy Misguided, Calls For its Withdrawal (dated Oct. 18, 2004), at http://www.discovery.org/scripts/ viewDB/index.php?command=view&id=2341 (last visited Jan. 27, 2006).

bers continued to find it difficult to define intelligent design or summarize its key tenets.[6]

Finally, although *Kitzmiller* was publicly portrayed as being about the "teaching" of intelligent design, in reality the Dover school board merely required students to hear a four-paragraph statement defining intelligent design as "an explanation of the origin of life that differs from Darwin's view"[7]—a vapid description that supplied virtually no meaningful information about the substance of the theory. Students were further notified that "[t]he reference book, *Of Pandas and People*, is available for students who might be interested in gaining an understanding of what Intelligent Design actually involves." Such a minimalist policy was a far cry from an intelligent design curriculum.

Nevertheless, the plaintiffs in *Kitzmiller* decided to turn their lawsuit into a broad referendum on the intellectual and scientific validity of intelligent design theory. Judge Jones eagerly obliged them, issuing a lengthy ruling purporting to analyze in detail the scientific claims made by ID proponents. Those who read Judge Jones' opinion without previously having studied the evidence and arguments presented in the case may well be impressed by Judge Jones' seemingly authoritative foray into the debate over evolution. They shouldn't be. As we will document, Judge Jones repeatedly misrepresented both the facts and the law in his opinion, sometimes egregiously (e.g., he asserted that scientists who support intelligent design have published no peer-reviewed articles or research, which is demonstrably false). When cross-checked against the evidence and arguments presented in the court record, many of Judge Jones' key assertions turn out to be erroneous, contradictory, or irrelevant.

The dogmatic tone of Judge Jones' opinion is already attracting criticism from thoughtful scholars. Distinguished University of Chicago

---

6. *Nightline: War In Dover* (ABC television broadcast, Jan. 13, 2005); Transcript of Testimony of Sheila Harkins 115-16, *Kitzmiller*, No. 4:04-CV-2688 (M.D. Pa., Nov. 2, 2005).

7. *Kitzmiller*, 2005 WL 3465563 at *4.

Law Professor Albert Alschuler, for one, has rebuked Judge Jones for smearing ID proponents as Biblical fundamentalists:

> If fundamentalism still means what it meant in the early twentieth century... accepting the Bible as literal truth—the champions of intelligent design are not fundamentalists. They uniformly disclaim reliance on the Book and focus only on where the biological evidence leads. The court's response—"well, that's what they say, but we know what they mean"—is uncivil, an illustration of the dismissive and contemptuous treatment that characterizes much contemporary discourse. Once we know who you are, we need not listen. We've heard it all already.[8]

According to Alschuler, in Judge Jones' eyes "Dover is simply Scopes trial redux. The proponents of intelligent design are guilty by association, and today's yahoos are merely yesterday's reincarnated."[9] Alschuler added that "proponents of intelligent design deserve the same respect" as evolutionists in the evaluation of their arguments, something they did not get from Judge Jones. Their ideas should be evaluated on their merits, not on presumed illicit motives. As Alschuler put it, "[f]reedom from psychoanalysis is a basic courtesy."

Even if Judge Jones' broad indictment of intelligent design had been exemplary, however, there is a serious question about whether it should have been issued at all.

It is a standard principle of good constitutional interpretation that a judge should venture only as far as necessary to answer the issue before him. If a judge can decide a case on narrow grounds, then that is what he ought to do.[10] He should not try to use his opinion to answer all possible

---

8. Albert Alschuler, The Dover Intelligent Design Decision, Part I: Of Motive, Effect, and History, The Faculty Blog University of Chicago Law School, (–Dec. 21, 2005), at http://uchicagolaw.typepad.com/faculty/2005/12/the_dover_intel.html (last visited Jan. 21, 2006).

9. *Id.*

10. One of the reasons for doing so is that the parties to a case typically have their own narrow interests, and may shade the presentation of the issues to

questions. In the present case, Judge Jones found that the Dover school board acted for clearly religious reasons rather than for a legitimate secular purpose.[11] Having made this determination, Supreme Court precedents required the conclusion that the policy adopted by the Dover board was unconstitutional.

That should have been the end of the decision. But Judge Jones decided to act as though it was the "intelligent design movement," not the Dover school board, that was on trial. Despite the fact that the lawyers who appeared in court on behalf of the Dover school board did not represent the "intelligent design movement," Judge Jones decided to launch an expansive inquiry into whether intelligent design is science, whether there is scientific evidence for the theory, whether the theory is inherently religious, whether Darwinism has flaws, and even whether Darwinian evolution is compatible with religious faith. A judge who practices judicial restraint would have recognized that those issues should only have been addressed in a case in which truly representative parties, with adequate notice of the issues at stake, had made the best arguments and presented the best evidence on both sides of each question.

Despite Judge Jones' protestations to the contrary, this *was* judicial activism—with a vengeance. Activist judges are not content for the normal processes in a democratic society to resolve divisive social controversies. Activist judges act as though the courts were specially delegated to substitute their "independent" (and absolutist) judgment for the political processes by which democratic societies resolve such controversies. Judges who espouse judicial restraint, by contrast, defer to the elected branches of government when they are presented with political questions, because it is ultimately the people themselves through the electoral pro-

---

favor those interests. If the judge sticks to the facts of the case before the court, and answers only those questions necessary to the resolution of the dispute at hand, then there will be opportunities in future cases for other interests and perspectives to be presented. The danger of pretending that the judge has all of the information and has heard the best arguments is reflected in *Kitzmiller*.

11. *Kitzmiller*, 2005 WL 3465563, at *35–*50.

cess who must be entrusted with deciding such questions.[12] In this case, Judge Jones did not need to answer any question beyond the question of whether the policy was religiously motivated. When he chose to venture beyond that question, to settle non-justiciable questions about the merits of the debate over intelligent design, Judge Jones succumbed to the same temptation of judicial activism that produced judicially-imposed "solutions" to a host of social conflicts during the second half of the twentieth century, such as abortion, capital punishment, prison reform, etc. Far from resolving controversial issues, such activism weakens both the political process and the authority of the courts, and often leads only to further social polarization. Unlike judicial activism, which seeks to decide issues by judicial fiat, the democratic process promotes incremental solutions and compromise, and it tends to cool passions over the long term by giving everyone a stake in the decision-making process.

Why, then, did Judge Jones venture so far afield from what was necessary to determine the case? From his comments to the newsmedia, it seems he yearned for his place in judicial history.[13] He relished the idea that he could be the first judge to issue a definitive pronouncement on ID, and he apparently was unwilling to forego that opportunity.

Judge Jones also had no small estimate of his own importance in the scheme of things. Take the remarkable passage from his opinion cited at the beginning of this Introduction. In it, Judge Jones boasts that "no other tribunal in the United States is in a better position than are we to traipse into this controversial area." He insists that his ruling on whether intelligent design is science "is essential" to his holding in the case, and is further motivated by his hope that he "may prevent the obvious waste of

---

12. *Baker v. Carr*, 369 U.S. 186, 210 (1962) ("The non-justiciability of a political question is primarily a function of the separation of powers").

13. See Bill Sulon, *'No Dover withdrawal for me,' intelligent-design trial judge says*, THE PATRIOT NEWS, Nov. 18, 2005, *at* http://www.pennlive.com/printer/printer.ssf?/base/news/1132322551160350.xml&coll=1&thispage=2 (last visited Jan. 30, 2006).

judicial and other resources which would be occasioned by a subsequent trial" on the issue of intelligent design.

This passage exhibits stunning presumption. First, and contrary to the Judge's claim, a determination of whether ID is science was plainly not essential to the disposition of the case, as has been pointed out.

Second, and more troubling, is the Judge's suggestion that his determination of whether ID is science would spare future judges the need to make their own determination. Judge Jones is a federal trial court judge in one particular district court in Pennsylvania. But he writes as if he has the right and duty to decide the question of whether intelligent design is science for all other judges in the entire United States in the future and, thereby, to legislate the question for the whole country. Lower federal court judges are bound by Supreme Court precedents, but they certainly aren't bound by the rulings of other lower court judges at the same level. Although other federal judges can refer to Judge Jones' decision (especially to his legal reasoning), every judge has a duty to reach an impartial and independent determination of the facts and law in the cases before him. Another federal district court judge would be remiss to simply say, "Well, Judge Jones has already decided the matter, so there is no need for me to do any fact-finding of my own." Nor should a judge tell the parties to a new case: "I've decided not to allow you to present any evidence, because Judge Jones already heard the evidence three years ago."

Whatever he may think to the contrary, Judge Jones has neither the authority nor the right to speak for the entire federal judiciary. Other judges will undoubtedly have the opportunity to revisit the issues examined in *Kitzmiller v. Dover*. When they do so, we hope they will ignore Judge Jones' one-sided analysis and reach their own conclusion. As we will show, *Kitzmiller* has little to contribute to the on-going dialogue about how to teach about biological evolution in the public schools, and it deserves no deference either from other jurists or from government officials. There are four major reasons for this:

1. ***Kitzmiller's* Partisan History of Intelligent Design.** Judge Jones purports to offer a definitive history of intelligent design as an off-shoot of "creationism," but the historical narrative he presents is shallow and one-sided, and suppresses many key facts.

2. ***Kitzmiller's* Unpersuasive Case against the Scientific Status of Intelligent Design.** The centerpiece of Judge Jones' opinion is his assertion that intelligent design is "is an interesting theological argument, but that it is not science." Not only does this assertion go well beyond the Judge's legitimate authority, it flatly contradicts both logic and the evidence presented in the court record.

3. ***Kitzmiller's* Failure to Treat Religion in a Neutral Manner.** Judges are required by the Constitution to treat religious questions neutrally, but Judge Jones applies different standards when examining the religious implications of intelligent design and Darwinian evolution. He even attempts to decide which theological view of evolution is correct.

4. ***Kitzmiller's* Limited Value as Precedent.** Judge Jones purports to answer once and for all the question of whether it is lawful to include intelligent design in public school science curricula, but in fact his opinion on this question has almost no precedential value for other judges.

Although the *Kitzmiller* decision does not deserve any deference because of its deep flaws, it already is being invoked by defenders of evolution to censor even voluntary expressions of disagreement with Darwinism. As the conclusion will point out, because of Judge Jones' complete failure to protect the academic freedom of teachers and students to express dissenting views on this topic, his opinion exposes the pressing need of other branches of government to move to protect this fundamental right.

CHAPTER I

KITZMILLER'S PARTISAN HISTORY OF INTELLIGENT DESIGN

A key part of Judge Jones' ruling is his purported history of the intelligent design movement, which he depicts as the outgrowth of American Christian "Fundamentalism" with a capital "F."[14] It is important to note that Judge Jones cannot point to even a single doctrine unique to Christian fundamentalism that the theory of intelligent design incorporates. Indeed, he effectively concedes that ID proponents distinguish their theory from fundamentalism by pointing out that it does *not* involve arguments based on "the Book of Genesis", "a young earth," or "a catastrophic Noaich flood."[15]

So where is the "Fundamentalism"? In conflating ID with fundamentalism, Judge Jones simply ignored the testimony in his court of two of the most prominent ID scientists, biologists Michael Behe and Scott Minnich. Neither Minnich nor Behe was shown by the plaintiffs to be a fundamentalist (they are not), neither was shown to believe in a literal reading of Genesis (they do not), neither was shown to come to his beliefs in ID from fundamentalism (they did not), and both reject neo-Darwinism on scientific grounds. Indeed, Behe has made clear that he had no problem with the modern theory of evolution until he discovered that what he was seeing in the lab did not fit with what he was being

---

14. *Kitzmiller*, 2005 WL 3465563 at *6–*11, *36.

15. *Id*. at *15–*16.

told in standard textbook accounts. Behe's skepticism of neo-Darwinism was not driven by a change in religion, but by scientific evidence. So again, where is the "Fundamentalism"? Judge Jones' use of the term "Fundamentalist" to describe intelligent design and the scientists who have propounded it is a descent into name-calling. Even news reporters hostile to intelligent design (and papers such as the *New York Times*) have recognized the error in applying this term to supporters of intelligent design; but not Judge John E. Jones.

Judge Jones' shallow and one-sided recital of the history of intelligent design owes less to the historical record than to the stereotypes of the old Hollywood film *Inherit the Wind*, whose accuracy as history has been debunked by scholars—but which Judge Jones told one reporter during the trial he planned to watch again for "historical context"![16] Most egregiously, Judge Jones' narrative completely ignores the long-standing and much broader debate over design in nature that has existed for millennia. As will be documented below, that broader debate over design predates the Bible, and the modern theory of intelligent design is not repackaged Biblical creationism, but rather a group of closely related arguments based on evidence from nature and our uniform experience of the cause-and-effect structure of the world.[17]

---

16. Amy Worden, "Bad Frog Beer to 'intelligent design': the controversial ex-Pa. liquor board chief is now U.S. judge in the closely watched trial," *Philadelphia Inquirer*, Oct. 16, 2005, at http://www.philly.com/mld/inquirer/news/local/states/pennsylvania/counties/montgomery_county/12912029.htm (last visited Nov. 6, 2005). For histories of the actual Scopes trial—fictionalized in *Inherit the Wind*—see Edward J. Larson, Summer for the Gods: the Scopes Trial and America's Continuing Debate over Science and Religion (New York, Basic Books 1997); Marvin Olasky and John Perry, Monkey Business: The True Story of the Scopes Trial (Nashville, Broadman and Holman 2005).

17. The narrative that follows draws on the amicus brief filed by the Foundation for Thought and Ethics.

A. The Ancient Origins of the Design Debate

Relying on theologian John Haught, Judge Jones treats Thomas Aquinas as the original source of the idea of intelligent design, writing that "ID is not a new scientific argument, but is rather an old religious argument for the existence of God. He [Haught] traced this argument back to at least Thomas Aquinas in the 13th century...."[18] Contra Judge Jones, the real origins of the debate over intelligent design reach back much further to ancient Greece and Rome. Greek philosophers Heraclitus, Empedocles, Democritus, and Anaximander proposed that life could originate without any intelligent guidance, while Socrates, Plato, and Aristotle advocated that mind was required.[19] During the Roman era, Cicero cited the orderly operation of the stars as well as biological adaptations in animals as empirical evidence that nature was the product of "rational design."[20]

Design was also an important part of the contemporary scientific debate at the time Darwin's theory was developed. Indeed, the term "intelligent design" as an alternative to blind evolution was employed by Oxford scholar F. C. S. Schiller as early as 1897. Schiller wrote that "it will not be possible to rule out the supposition that the process of Evolution may be guided by an intelligent design."[21] Schiller, like modern

---

18. *Kitzmiller*, 2005 WL 3465563, at *12.

19. *See* XENOPHON, MEMORABILIA OF SOCRATES, Book I, chapter 4; PLATO, THE LAWS, Book X; Michael Ruse, "The Argument from Design: A Brief History," in *Debating Design* 13-16 (William A. Dembski & Michael Ruse eds., Cambridge, Cambridge University Press 2004); John Angus Campbell, "Why Are We Still Debating Darwinism? Why Not Teach the Controversy?" *in Darwin, Design, and Public Education* xii (John Angus Campbell & Stephen C. Meyer eds., East Lansing, Michigan State University Press 2003).

20. Cicero, *De Natura Deorum* 217, 237, 245 (H. Rackham, trans., Cambridge, Harvard University Press 1933).

21. F. C. S. Schiller, *Darwinism and Design Argument*, in HUMANISM: PHILOSOPHICAL ESSAYS 141 (F. C. S. Schiller, New York, The Macmillan Co. 1903). This particular essay was first published in the *Contemporary Review* in June, 1897.

design theorist Michael Behe, argued for intelligent design without rejecting all forms of evolution or even common descent.

Prominent nineteenth century scientists held similar views, including even Alfred Russel Wallace, the co-developer with Charles Darwin of the theory of evolution by natural selection. By the late nineteenth century, Wallace came to believe that natural selection acting on random variations could not explain a number of things in biology, especially the development of the human brain. He concluded instead that "a Higher Intelligence" guided the process:

> [T]here seems to be evidence of a Power which has guided the action of those laws [of organic development] in definite directions and for special ends. And so far from this view being out of harmony with the teachings of science, it has a striking analogy with what is now taking place in the world....[22]

While Wallace certainly ascribed more religious meaning to his concept than was warranted by the data, he nonetheless recognized that it was possible to detect design in nature. It is ironic that Judge Jones' decision effectively renders unconstitutional the views of the co-founder of the modern theory of evolution.

B. The Modern Revival of the Design Debate in Physics and Cosmology

Although Judge Jones' history of intelligent design is preoccupied with biology, the revival of design in science in the twentieth century did not originate in biology. It started in physics, astronomy, and cosmology. Beginning with Fred Hoyle's discovery of the carbon-12 resonance in the early 1950s,[23] physicists uncovered a number of ways the universal

22. Alfred Russel Wallace, *Sir Charles Lyell on Geological Climates the Origin of Species, in* An Anthology of His Shorter Writings 33–34 (Charles H. Smith ed., Oxford University Press 1991).

23. Fred Hoyle, *On Nuclear Reactions Occurring in Very Hot Stars. I. The Synthesis of Elements from Carbon to Nickel,* 1 Astrophysical Journal Supplement 121–46 (1954).

constants of physics and chemistry (gravity, the strong and weak nuclear forces, etc.) were fine-tuned for complex life. Reviewing these developments in 1982, the noted theoretical physicist Paul Davies described the fine-tuning of the universe as "the most compelling evidence for an element of cosmic design."[24]

Hoyle, an eminent theoretical physicist and agnostic, followed with *The Intelligent Universe* (1983), featuring chapter titles like "The Information Rich Universe" and "What is Intelligence Up To?" Hoyle, no friend of Christianity or Biblical creationism, nevertheless asserted, "A component has evidently been missing from cosmological studies. The origin of the Universe, like the solution of the Rubik cube, requires an intelligence."[25] Or as Hoyle said elsewhere, "A commonsense interpretation of the facts suggests that a superintellect has monkeyed with physics, as well as chemistry and biology, and that there are no blind forces worth speaking about in nature."[26]

Hoyle's argument, thus, extended even to the realm of biology: "We are close here to what seems to be going on in the mind of the Darwinian enthusiast, whose processes of thought seem to be conditioned by the tacit assumption that the environment is intelligent, an idea which I would in part subscribe to, but one which in Darwinian theory is quite against the rules." Hoyle added that "[a] proper understanding of evolution requires that the environment, or the variations on which it operates, or both, be intelligently controlled."[27] Other scientists in biology, chemistry, and related disciplines were already elaborating on this theme as they sought to better understand the continued mystery of biological origins.

---

24. PAUL DAVIES, THE ACCIDENTAL UNIVERSE 189 (Cambridge, Cambridge University Press, 1982).

25. FRED HOYLE, THE INTELLIGENT UNIVERSE 189 (New York, Holt, Rinehart, and Winston 1983).

26. Fred Hoyle *quoted in* DAVIES, *supra* note 24, at 118.

27. HOYLE, *supra* note 25, at 244.

C. The Modern Revival of the Design Debate in Biology

Relying on polemical ID critics such as the philosopher Barbara Forrest, Judge Jones depicts the intelligent design movement in biology as an effort to repackage Biblical creationism in order to circumvent the Supreme Court's 1987 decision in *Edwards v. Aguillard*, which struck down a Louisiana creationism law as violative of the Establishment Clause. However, in his book *By Design*, journalist Larry Witham traces the roots of the intelligent design movement in biology to the 1950s and 1960s, and the movement itself to the 1970s.[28] Biochemists were unraveling the secret of DNA and discovering that it was part of an elaborate information processing system that included nanotechnology of unparalleled sophistication. One of the first intellectuals to describe the significance of these discoveries was chemist and philosopher Michael Polanyi, who in 1967 argued that "machines are irreducible to physics and chemistry" and that "mechanistic structures of living beings appear to be likewise irreducible."[29]

Biochemist Michael Behe would later develop Polanyi's insights with his concept of irreducible complexity. And mathematician William Dembski would find Polanyi's work so influential, that he would name Baylor University's Michael Polanyi Center after him.

Polanyi's work also influenced the seminal 1984 book *The Mystery of Life's Origin* by Charles Thaxton (Ph.D., Physical Chemistry, Iowa State University), Walter Bradley (Ph.D., Materials Science, University of Texas, Austin), and Roger Olsen (Ph.D., Geochemistry, Colorado School of Mines). Thaxton and his co-authors argued that matter and energy can accomplish only so much by themselves, and that some things can only "be accomplished through what Michael Polanyi has

---

28. LARRY WITHAM, BY DESIGN (San Francisco, Encounter Books 2003).

29. Michael Polanyi, *Life transcending physics and chemistry*, 45 (35) CHEMICAL AND ENGINEERING NEWS 54–66 (Aug. 21, 1967).

called 'a profoundly informative intervention.'"[30] The book was placed
with The Philosophical Library of New York, publisher of more than
twenty Nobel laureates, and became the best-selling advanced college-
level work on chemical evolution. Sales were fueled by favorable reviews
in prestigious venues like the *Yale Journal of Biology and Medicine*, as well
as a positive response from leading scholars.[31]

In the same year *The Mystery of Life's Origin* appeared, Thaxton met
Stephen Meyer, a young geophysicist and future philosopher of science.
Meyer was to become a founder and director of Discovery Institute's
Center for Science & Culture, now the institutional home for scientists
and scholars around the globe working on the theory of intelligent design. Thaxton, Meyer, and others were by the mid-1980s already using
terms like *creative intelligence, intelligent cause, artificer,* and *intelligent artificer,* as they grappled together with questions of design detection in
science. As a consequence, the basic concepts of what became known as
intelligent design appear in the work of Thaxton and others well before
the Supreme Court handed down its 1987 decision concerning creationism in *Edwards v. Aguillard.*

As the academic editor for the Foundation of Thought and Ethics,
Thaxton was then serving as the editor for a supplemental science textbook co-authored by San Francisco State University biologist Dean Kenyon and eventually named *Of Pandas and People: The Central Question
of Biological Origins.* As that book neared completion, Thaxton continued to pitch around for an overarching term that was less ponderous and,
at the same time, more general than the current options before them, a
term to describe a science open to evidence for intelligent causation and
free of religious assumptions. He found it in a phrase he picked up from
a NASA scientist, intelligent design. "That's just what I need," Thaxton

---

30. CHARLES B. THAXTON ET AL., THE MYSTERY OF LIFE'S ORIGIN 185 (Dallas, Lewis and Stanley 1984).

31. For example, it received approval from Klaus Dose in his major review article on origin of life studies. See Klaus Dose, *The Origin of Life: More Questions
Than Answers,* 13 (4) INTERDISCIPLINARY SCIENCE REVIEWS (1988).

recalls thinking.[32] "It's a good engineering term...." It was soon incorporated into the language of the book.

Judge Jones makes much of the fact that certain early drafts of *Pandas* used the terms "creation" and "creationists." He concludes that the substitution of "intelligent design" for "creation" or "creationist" was in order to evade the legal effect of the decision in *Edwards v. Aguillard*. While it is undoubtedly true that the decision in *Edwards* affected subsequent editorial decisions, Judge Jones drew precisely the wrong implication. It was not that the Supreme Court in *Edwards* had decided that it was unconstitutional to teach the concept that was being advanced in *Pandas*; rather, it was that the Supreme Court had used "creationism" to mean something very different from what the authors of *Pandas* were trying to communicate, and they wanted to be sure that no one confused the ideas they were advancing for the "creationism" that was the subject of the Louisiana statute.

Long before *Edwards*, the authors of *Pandas* specifically rejected the view that science could detect whether the intelligent cause identified was supernatural. Of course, the process by which an intelligent agent produces a designed object might loosely be called a "creation" (as in stating that this article was the "creation" of several authors), and the movement attributing apparent design in nature might in this sense be described as "creationist." But defining "creationism" in such an overbroad manner would mean that theistic evolutionists such as Kenneth Miller are also "creationists," as Miller himself acknowledged on the stand:

Q. Sir, in the ordinary meaning of the word a creationist is simply any person who believes in an act of creation, correct?

B. Yes, I think I would also regard that as the ordinary meaning of the word creationist.

Q. And you believe that the universe was created by God?

---

32. Jonathan Witt, *The Origins of the Term Intelligent Design*, at http://www. discovery.org/scripts/viewDB/filesDB-download.php?command=download &id=526 (last visited Jan. 30, 2006).

A. I believe that God is the author of all things seen and unseen. So the answer to that, sir, is yes.

Q. In a sense that would make you a creationist using the definition—

A. ... in that sense any person who is a theist, any person who accepts a supreme being, is a creationist in the ordinary meaning of the word because they believe in some sort of a creation event.

Q. And that would include yourself?

A. That would certainly include me.[33]

If Miller's admission that he is a "creationist" in the "ordinary meaning of the word" does not make him an advocate of "creationism" as that term is generally understood today, why should the authors of *Pandas* be held to a different standard? Early drafts of *Pandas*, whatever their variations in terminology, clearly did not advocate what is generally understood as "creationism." Indeed, a pre-*Edwards v. Aguillard* draft from the first part of 1987 emphatically stated that "observable instances of information cannot tell us if the intellect behind them is natural or supernatural. This is not a question that science can answer."[34] The same draft rejected the eighteenth-century design argument from William Paley because it illegitimately tried "to extrapolate to the supernatural" from the empirical data of science. Paley was wrong because "there is no basis in uniform experience for going from nature to the supernatural, for inferring an unobserved supernatural cause from an observed effect."[35] Similarly, another early draft (also from when the manuscript was still titled "Biology and Origins") stated:

---

33. Transcript of Testimony of Kenneth Miller 62–3, *Kitzmiller*, No. 4:04-CV-2688 (M.D. Pa., Sept. 27, 2005).

34. Charles Thaxton, *Introduction to Teachers*, in Dean H. Kenyon & P. William Davis, BIOLOGY AND ORIGINS 13 (Foundation for Thought and Ethics, Manuscript # I 1987).

35. *Id.*

[T]here are two things about which we cannot learn through uniform sensory experience. One is the supernatural, and so to teach it in science classes would be out of place.... [S]cience can identify an intellect, but is powerless to tell us if that intellect is within the universe or beyond it.[36]

By unequivocally affirming that the empirical evidence of science "cannot tell us if the intellect behind [the information in life] was natural or supernatural"[37] it is evident that these early drafts of *Pandas* meant something very different by "creation" than did the Supreme Court in *Edwards v. Aguillard*, which defined creationism as requiring a "supernatural creator."[38] The decision to use the term "intelligent design" in later drafts to express the book's central idea was not an attempt to evade a court decision, but rather to furnish a more precise description of the emerging scientific theory. As Thaxton stated in his deposition for the *Kitzmiller* case:

I wasn't comfortable with the typical vocabulary that for the most part creationists were using *because it didn't express what I was trying to do. They were wanting to bring God into the discussion, and I was wanting to stay within the empirical domain and do what you can do legitimately there.*[39]

Judge Jones was informed of the broader historical context of the debate over design by the amicus brief that was submitted by the Foundation for Thought and Ethics, but he apparently was not interested, preferring instead the ACLU's false and tendentious history of the design movement.

---

36. *Id.* at 7–8.

37. *Id.* at 13.

38. *Edwards*, 482 U.S. at 592.

39. Deposition of Charles Thaxton, 52–53, *Kitzmiller*, No. 4:04-CV-2688 (M.D. Pa., July 19, 2005) (emphasis added).

CHAPTER II

KITZMILLER'S UNPERSUASIVE CASE AGAINST THE SCIENTIFIC STATUS OF INTELLIGENT DESIGN

The heart of Judge Jones' opinion is his extended discussion of the scientific status of intelligent design. In a finding trumpeted by defenders of Darwin's theory, Judge Jones asserted that over the duration of the trial he had reached "the inescapable conclusion that ID is an interesting theological argument, but that it is not science."[40] The number of mistakes Judge Jones made in reaching this "inescapable" conclusion are so numerous it is difficult to catalogue them all. However, they can be summarized as follows: (A) Judge Jones wrongly assumed the authority to decide what science is; (B) Judge Jones conflated the question of *whether* something is scientific with the question of *which scientific theory is most popular*; and (C) Judge Jones disqualified ID as science only by repeatedly misrepresenting the facts.

A. Judge Jones Wrongly Assumed the Authority To Decide What Science Is

The boundaries of science are not established by science itself but by philosophy, and the fascinating question of what constitutes science has vexed philosophers of science for many years. The last time a federal

---

40. *Kitzmiller*, 2005 WL 3465563 at *35.

judge attempted to define science for purposes of determining the constitutionality of teaching theories other than Darwinian evolution, the philosophy of science community responded critically.[41] Scientific methods and scientific theories cover such a broad sweep that it is difficult to provide a precise and consistent definition. As philosopher of science Larry Laudan explains, "If we would stand up and be counted on the side of reason, we ought to drop terms like 'pseudo-science'.... [T]hey... do only emotive work for us."[42] Or as Martin Eger has summarized, "[d]emarcation arguments have collapsed. Philosophers of science don't hold them anymore. They may still enjoy acceptance in the popular world, but that's a different world."[43] Elsewhere Laudan summarizes the state of the question in his field by commenting: "[T]here is no demarcation line between science and non-science, or between science and pseudo-science, which would win assent from a majority of philosophers."[44] This disagreement among philosophers as to the nature of science indicates "a lack of judicially discoverable and manageable standards."[45] If expert philosophers of science have been unable to settle the general question "what is science?," then Judge Jones had no judicial basis for finding that intelligent design is not science. In short, the nature of science should have been treated as a nonjusticiable issue by Judge Jones.

To be sure, federal judges regularly decide what is admissible or inadmissible as scientific testimony, and in doing so they must make determinations about whether an expert's testimony concerns scientific knowledge that will be helpful to the trier of fact in resolving the

---

41. See *McLean v. Arkansas Bd. of Education* 529 F. Supp. 1255, 1267 (E.D. Ark. 1982). *See also* the critical response of the philosophy of science community in But is it Science? (Michael Ruse, ed., Buffalo, Prometheus Books 1988).

42. Larry Laudan, *The Demise of the Demarcation Problem, in* But Is It Science?, *supra* note 41, at 337, 349.

43. John Buell, *Broaden Science Curriculum,* Dallas Morning News, March 10, 1989, at A21 (quoting an unidentified "authority").

44. Larry Laudan, Beyond Positivism and Relativism 210 (Boulder, Westview Press 1996).

45. *Vieth v. Jubelirer,* 541 U.S. 267, 277–78 (2004).

contested issues in a trial.[46] For example, a party may offer an expert to testify about how an accident happened or whether or not a disease was caused by exposure to the defendant's product. Before admitting the testimony the trial judge must determine the scientific reliability of the testimony being offered. Yet in doing so, the Supreme Court made it clear that the trial judge must be aware of the difference between the need to resolve particular disputes and the wide-ranging and tentative nature of scientific inquiry.[47]

A judge who presides over a case involving a scientific dispute must be careful not to confuse the need for a resolution—a "choosing of sides"—with a resolution of the scientific controversy itself. The continuing debate over the merits of the scientific controversy should be left unhampered by the judge's responsibility to take sides. Still less should a judge arrogate to himself the responsibility of deciding what constitutes science; that is a far more difficult question that should be left to scientists and scholars, not the courts.

A group of 85 scientists filed an amicus brief with Judge Jones making precisely this plea: "Amici strenuously object to appeals to the judiciary to rule on the validity of a scientific theory or to rule on the scope of science in a manner that might exclude certain scientific theories from science. These questions should be decided by scientists, not lawyers."[48]

---

46. *Daubert v. Merrell Dow Pharmaceuticals, Inc.*, 509 U.S. 579 (1993).

47. "[T]here are important differences between the quest for truth in the courtroom and the quest for truth in the laboratory. Scientific conclusions are subject to perpetual revision. Law, on the other hand, must resolve disputes finally and quickly. The scientific project is advanced by broad and wide-ranging consideration of a multitude of hypotheses, for those that are incorrect will eventually be shown to be so, and that in itself is an advance. Conjectures that are probably wrong are of little use, however, in the project of reaching a quick, final, and binding legal judgment... about a particular set of events in the past." *Daubert*, 509 U.S. at 596–97.

48. Brief of Amici Curiae Biologists And Other Scientists In support of Defendants at 7, *Kitzmiller* (No. 4:04-cv-2688), *at* http://www.discovery.org/scripts/viewDB/filesDB-download.php?command=download&id=558 (last visited Jan. 30, 2006). This brief is reprinted as Appendix C.

The scientists who made this appeal included a member of the National Academy of Sciences, a Fellow of the American Association for the Advancement of Science, and biologists and other scientists at such institutions as the University of Michigan, the University of Georgia, and the University of Wisconsin. Judge Jones did not even attempt to respond to their argument. Indeed, he did not even acknowledge its existence.

B. Judge Jones Conflated the Question of Whether Something is Scientific with the Question of Which Scientific Theory is Most Popular

Even if one were to grant, *ad arguendo,* that Judge Jones had the right to ascertain whether intelligent design is a scientific theory, that does not mean he did so in an intellectually credible way. As will be discussed below, much of Judge Jones' analysis of intelligent design is based on straw-man arguments and serious errors of fact. But there is a deeper theoretical problem with his analysis. Judge Jones does not appear to grasp that the question of *whether* a particular theory is scientific is distinct from the question of *which scientific theory is most popular among scientists.* For example, there are many competing scientific explanations for the origin of the first life. The fact that some of these explanations are not as strongly supported by the community of origin-of-life scientists as others does not make the less-supported explanations unscientific in principle. Yet Judge Jones repeatedly implies that "ID is not science" merely because the majority of scientists continue to believe in Darwinian evolution. By that standard, no new theory proposed by scientists would ever be considered "scientific" until it became dominant in the scientific community. Such an artificially-constrained definition of science has never served as a common standard in the scientific community and, if it ever did, would have a chilling effect on scientific research and debate. Science could never correct its errors if challenges to the dominant point of view were immediately labeled "unscientific."

C. Judge Jones Disqualified ID as Science Only by Misrepresenting the Facts

Despite Judge Jones' grand aspiration to make his opinion the "gold standard" for future considerations of this issue, he offered surprisingly weak evidence for many key factual assertions in the case. Much of his criticism of the theory of intelligent design responds to the testimony of Lehigh University biochemist Michael Behe. But Behe has written a point-by-point answer to each of Judge Jones' claims, showing that Judge Jones fundamentally misrepresented Behe's testimony. Readers are urged to review Behe's detailed response, which is included in this book as Appendix A.

In addition, we can examine Judge Jones' own summary of his reasons why intelligent design is not science. On page 62 of his opinion, Judge Jones identifies six claims which were relevant to his determination:

> (1) ID violates the centuries-old ground rules of science by invoking and permitting supernatural causation; (2) the argument of irreducible complexity, central to ID, employs the same flawed and illogical contrived dualism that doomed creation science in the 1980's; and (3) ID's negative attacks on evolution have been refuted by the scientific community... it is additionally important to note that [4] ID has failed to gain acceptance in the scientific community, [5] it has not generated peer-reviewed publications, nor [6] has it been the subject of testing and research.[49]

As will be discussed in detail below, each of the above claims is either false or irrelevant.

---

49. *Kitzmiller*, 2005 WL 3465563, at *26.

1. "ID violates the centuries-old ground rules of science by invoking and permitting supernatural causation."

Judge Jones makes two interrelated claims here that need to be distinguished: a. ID invokes or permits supernatural causation and b. ID violates the centuries-old ground rules of science.

a. Does ID invoke or permit supernatural causation?

Although Judge Jones sometimes claims that ID *either* "invokes *or* permits supernatural causation," it becomes clear in his opinion that his real claim is much stronger: He repeatedly insists that ID "requires supernatural creation."[50] Judge Jones can make this claim only by misrepresenting the actual views of intelligent design scientists, who consistently have maintained that empirical evidence cannot tell one whether the intelligent causes detected through modern science are inside or outside of nature. As a scientific theory, ID only claims that there is empirical evidence that key features of the universe and living things are the products of an intelligent cause. Whether the intelligent cause involved is inside or outside of nature cannot be decided by empirical evidence alone. That larger question involves philosophy, including metaphysics. In addition to the clear testimony of ID witnesses during the trial on this point, Judge Jones was provided with fifteen pages of documentation unequivocally demonstrating that ID proponents from the beginning have repeatedly argued that design theory does not rely on supernatural causation, and they have consistently maintained this position whether writing for religious or secular audiences.[51] Ignoring this evidence, Judge Jones proceeded to highlight a few quotations cited by the plaintiffs to prove his conclusion that ID requires supernatural causation. However, Judge Jones distorts the plain meaning of these quotations, which contradict, rather than support, his claim.

---

50. *Id.* at 14.

51. Brief (Revised) of Amicus Curiae Discovery Institute app. at 1–15, *Kitzmiller* (No. 4:04-cv-2688), *at* http://www.discovery.org/scripts/viewDB/filesDB-download.php?command=download&id=647 (last visited Jan. 30, 2006).

For example, Judge Jones argues that the *Pandas* textbook "indicates that there are two kinds of causes, natural and intelligent, *which demonstrate that intelligent causes are beyond nature.*"[52] Yet *Pandas* explicitly and repeatedly makes precisely the *opposite* claim: Intelligent causes may be either inside or outside of nature, and empirical evidence alone cannot determine which option is correct. *Pandas* made this distinction even in an early pre-publication draft which emphatically stated that "in science, the proper contrary to natural cause is not supernatural cause, but intelligent cause."[53] Also consider the following passage from the edition of *Pandas* actually used in the Dover school district:

> If science is based upon experience, then science tells us the message encoded in DNA must have originated from an intelligent cause. *But what kind of intelligent agent was it? On its own, science cannot answer this question; it must leave it to religion and philosophy.* But that should not prevent science from acknowledging evidences for an intelligent cause origin wherever they may exist.[54]

Incredibly, Judge Jones misinterprets the above passage as further proof that ID requires a belief in a supernatural cause, claiming:

> In fact, an explicit concession that the intelligent designer works outside the laws of nature and science and a direct reference to religion is *Pandas'* rhetorical statement, "what kind of intelligent agent was it [the designer]" and answer: "On its own science cannot answer this question. It must leave it to religion and philosophy."[55]

Contrary to Judge Jones, the above statement clearly does *not* concede that "the intelligent designer works outside the laws of nature and science." Instead, it merely reaffirms that empirical science cannot de-

---

52. *Kitzmiller*, 2005 WL 3465563, at *14 (emphasis added).

53. Brief of Amicus Curiae FTE appendix B at 6, *Kitzmiller* (No. 4:04-cv-2688), *at* http://www.discovery.org/scripts/viewDB/filesDB-download.php?command=download&id=648 (last visited Jan. 30, 2006).

54. Dean H. Kenyon and Percival Davis, Of Pandas and People 7 (2nd ed. Foundation for Thought and Ethics 1993) (emphasis added).

55. *Kitzmiller*, 2005 WL 3465563, at *12.

termine whether the intelligent cause detected resides inside or outside of nature. That further determination requires more than empirical science. Far from being merely "rhetorical," this claim is central to the definition of intelligent design as a scientific theory, and it is reaffirmed and further explained in other passages in *Pandas* that the Judge ignores.[56]

Judge Jones' distorted analysis of the views of ID proponents extends to the ID expert witnesses, Michael Behe and Scott Minnich. In his opinion, Judge Jones cites a comment by Michael Behe that intelligent design means "designed not by the laws of nature" as evidence that "a supernatural designer is a hallmark of ID."[57] But is that what Behe meant? Consider the full context of this quote: "They were designed not by the laws of nature, not by chance and necessity; rather they were planned."[58] Thus Behe notes that design means that the laws of nature—i.e. chance and necessity—are ruled out because an intelligent agent acted. This is different from asserting that the intelligence involved must be supernatural. Indeed, under Behe's account, even a human intelligence can plan something and act outside of the chance and necessity which characterize the laws of nature. But a human intelligent agent is not "supernatural." Since our basis for inferring intelligent design generally is found in our observations of how we observe human intelligence acting,[59] it is

---

56. "Today we recognize that appeals to intelligent design may be considered in science, as illustrated by current NASA search for extraterrestrial intelligence (SETI). Archaeology has pioneered the development of methods for distinguishing the effects of natural and intelligent causes. *We should recognize, however, that if we go further, and conclude that the intelligence responsible for biological origins is outside the universe (supernatural) or within it, we do so without the help of science.*" Dean H. Kenyon and Percival Davis, *supra* note 54, at 126–27 (emphasis added).

57. *Kitzmiller*, 2005 WL 3465563, at *14. The original quotation is from MICHAEL BEHE, DARWIN'S BLACK BOX 193. (New York, Free Press 1996). Judge Jones' decision misquotes this passage slightly as "not designed by the laws of nature."

58. Behe, *supra* note 57, at 193.

59. Stephen C. Meyer, *The Origin of Biological Information and the Higher Taxonomic Categories*, 117 (2) PROCEEDINGS OF THE BIOLOGICAL SOCIETY OF

clear that the scientific theory of intelligent design does not require the intelligent causes it detects to be outside nature.

Contrary to the interpretation of Judge Jones, Behe reiterates throughout both his writings and his court testimony that intelligent design does *not* require a supernatural entity:

> The conclusion that something was designed can be made quite independently of knowledge of the designer. As a matter of procedure, the design must first be apprehended before there can be any further question about the designer. The inference to design can be held with all the firmness that is possible in this world, without knowing anything about the designer.[60]

Scott Minnich also made clear during his trial testimony that intelligent design does not require a supernatural designer:

Q. Do you have an opinion as to whether intelligent design requires the action of a supernatural creator?

A. I do.

Q. What is that opinion?

A. It does not.[61]

Because there was no genuine evidence that the theory of intelligent design requires a supernatural cause, Judge Jones had to fall back on quotes from intelligent design proponents expressing their personal or philosophical beliefs about God.[62] But the personal religious beliefs of ID proponents ought to be irrelevant. The issue is not whether Michael Behe or Scott Minnich—like the vast majority of Americans—believe in God, or that they believe (based on their personal religious beliefs)

WASHINGTON, 213-229 (2004).

60. Behe, *supra* note 57 at 197; Transcript of Testimony of Michael Behe 90, *Kitzmiller*, No. 4:04-CV-2688 (M.D. Pa., Oct. 18, 2005).

61. Transcript of Testimony of Scott Minnich 45-6, 135, *Kitzmiller*, No. 4:04--CV-2688 (M.D. Pa., Nov. 3, 2005).

62. *Kitzmiller*, 2005 WL 3465563, at *12–*14.

that the intelligent agent identified in their research is the God of the Bible. The issue is whether they claim that science must be based on such a belief. Clearly, they do not. If the personal metaphysical beliefs of a theory's proponents, or the metaphysical conclusions to which their theories lead them, were enough to make a theory religious rather than scientific, then the leading proponents of neo-Darwinism such as Richard Dawkins and Daniel Dennett would clearly be disqualified as scientists. Subjecting the personal religious beliefs of ID proponents to microscopic examination, while ignoring the beliefs of Darwinian scientists, was one of the more egregious examples of the double standard that Judge Jones adopted throughout the trial. This double standard will be discussed in detail in Chapter III.

While intelligent design does not require a supernatural intelligence, Judge Jones is correct that the theory allows for the *possibility* that the intelligent cause detected through science is supernatural. So what? Unless Judge Jones believes that science must be explicitly atheistic, why is it a problem that ID may *allow* for the *possibility* that an intelligent cause is supernatural? After all, theistic evolutionists such as Kenneth Miller believe that the evolutionary process allows for the action of a supernatural agent. Does that make their view of evolution religion rather than science? Given his own later statements about the compatibility of religion and evolution, Judge Jones obviously does not think so.[63] Why, then, does he insist that ID cannot permit the possibility of a supernatural cause without being placed outside the boundaries of science?

b. Does ID violate "the centuries-old ground rules of science"?

Judge Jones further faults ID for violating "methodological naturalism" (sometimes called "methodological materialism"[64]), which he re-

---

63. *Kitzmiller*, 2005 WL 3465563, at *51.

64. "This restriction of evolution to explanation through natural cause is referred to as 'methodological materialism,' materialism in this context referring to matter, energy, and their interaction. Methodological materialism is one of the main differences between science and religion," Eugenie C. Scott, *Science,*

peatedly credits as a foundational "ground rule" of science for the past several centuries. Methodological naturalism "limits inquiry to testable, natural explanations about the natural world."[65] Whether methodological naturalism is really a foundational ground rule for the operation of science has been sharply disputed by historians and philosophers of science.[66] Assuming *ad arguendo* that Judge Jones is correct, his argument proves far less than he believes. Intelligent design, properly conceived, does not need to violate methodological naturalism, a point that expert witness Scott Minnich made clear at trial.[67]

To understand why this is the case, one needs to understand how a design inference is drawn. Intelligent design theory assumes that intelligence is a property which we can understand through general observation of intelligent agents in the natural world. An intelligent agent exhibits predictable modes of designing because it has the property of intelligence, regardless of whether or not the agent is "natural" or "supernatural." Thus, the theory of intelligent design does not investigate whether the designing intelligent agent was natural or supernatural because it assumes that things designed by an intelligence may possess certain perceptible properties regardless of whether that intelligent agent is a natural entity, or in some way supernatural.

Contrary to Judge Jones, intelligent design is clearly based upon an explanatory cause whose behavior is understandable and yields predictable evidence that it was at work. Mathematician and philosopher William Dembski has observed that "[t]he principal characteristic of

---

Religion and Evolution, at http://www.ncseweb.org/resources/articles/6366_science_religion_and_evoluti_6_19_2001.asp (last visited Jan. 26, 2005).

65. *Kitzmiller,* 2005 WL 3465563, at *26.

66. Larry Laudan, *Science at the Bar—Causes for Concern, in* BUT IS IT SCIENCE? *supra* note 41 at 351, 355.

67. Transcript of Testimony of Scott Minnich 137, *Kitzmiller,* No. 4:04-CV-2688 (M.D. Pa., Nov. 3, 2005).

intelligent agency is directed contingency, or what we call choice."[68] By observing the sorts of choices that intelligent agents commonly make when designing systems, design theorists can identify reliable indicators of when an intelligent agent was involved in the origin of an object.

Intelligent design thus begins with observations about how intelligent agents operate. It then proceeds to convert those observations into predictions of what we might find if intelligent design was involved in the origin of a given natural object. For example, Stephen C. Meyer observes the following about how intelligent agents operate:

> Agents can arrange matter with distant goals in mind. In their use of language, they routinely "find" highly isolated and improbable functional sequences amid vast spaces of combinatorial possibilities.[69]

> [W]e have repeated experience of rational and conscious agents—in particular ourselves—generating or causing increases in complex specified information, both in the form of sequence-specific lines of code and in the form of hierarchically arranged systems of parts. Our experience-based knowledge of information-flow confirms that systems with large amounts of specified complexity (especially codes and languages) invariably originate from an intelligent source from a mind or personal agent.[70]

After identifying such "signs of intelligence," Meyer explains that we have scientific justification for invoking intelligent causes:

> [B]y invoking design to explain the origin of new biological information, contemporary design theorists are not positing an arbitrary explanatory element unmotivated by a consideration of the evidence. Instead, they are positing an entity possessing precisely

---

68. William A. Dembski, The Design Inference 62 (Cambridge, Cambridge University Press 1998).

69. Stephen C. Meyer, *The Cambrian Information Explosion, in* Debating Design *supra* note 19, at 388.

70. Meyer, *supra* note 59.

the attributes and causal powers that the phenomenon in question requires as a condition of its production and explanation.[71]

Note that throughout this process, the only explanations invoked are based upon evidence in nature, including our observational experience about how intelligent agents operate.

This method fits precisely with how the National Academy of Sciences characterizes science, praised by the Judge on page 66 of his opinion:

> Science is a particular way of knowing about the world. In science, explanations are restricted to those that can be inferred from the confirmable data—the results obtained through observations and experiments that can be substantiated by other scientists. Anything that can be observed or measured is amenable to scientific investigation. Explanations that cannot be based upon empirical evidence are not part of science.[72]

Intelligent causes can be inferred through confirmable data. The types of information produced by intelligent causes can be observed and then measured. Scientists can use observations and experiments to base their conclusions of intelligent design upon empirical evidence. Intelligent design limits its claims to those which can be established through the data. In this way, intelligent design does not violate the mandates of predictability and reliability laid down for science by methodological naturalism (whatever the failings and limitations of methodological naturalism).

---

71. *Id.*

72. NATIONAL ACADEMY OF SCIENCES, TEACHING ABOUT EVOLUTION AND THE NATURE OF SCIENCE 27 (Washington, D.C., National Academy Press 1998).

2. "The argument of irreducible complexity, central to ID, employs the same flawed and illogical contrived dualism that doomed creation science in the 1980's."

According to Judge Jones, the use of irreducible complexity as a demonstration of intelligent design rests on a "contrived dualism," because it falsely claims that if "evolutionary theory is discredited, ID is confirmed."[73] Yet on closer inspection it is Judge Jones' charge of "contrived dualism" that is truly contrived. The way Michael Behe formulates his theory, neo-Darwinism predicts we will not find irreducible complexity, while intelligent design predicts we will. This is not a form of "contrived dualism." This is an actual, logical dualism.

Why irreducible complexity is a negative prediction of evolution. Michael Behe has clearly explained why irreducible complexity provides negative evidence against neo-Darwinism:

> In the *Origin of the Species,* Charles Darwin said, "If it could be demonstrated that any complex organ existed which could not possibly have been formed by numerous, successive, slight modifications, my theory would absolutely break down." A system which meets Darwin's criterion is one which exhibits irreducible complexity. By irreducible complexity I mean a single system which is composed of several interacting parts that contribute to the basic function, and where the removal of any one of the parts causes the system to effectively cease functioning. An irreducibly complex system cannot be produced gradually by slight, successive modifications of a precursor system, since any precursor to an irreducibly complex system is by definition nonfunctional. Since natural selection requires a function to select, an irreducibly complex biological system, if there is such a thing, would have to arise as an integrated unit for natural selection to have anything to act on. It is almost universally conceded that

---

73. *Kitzmiller,* 2005 WL 3465563, at *28.

such a sudden event would be irreconcilable with the gradualism Darwin envisioned.[74]

Thus Darwinian gradual selection cannot produce irreducibly complex structures as Behe has described them. As a result, irreducible complexity supplies negative evidence against the sufficiency of the Darwinian mechanism to build certain kinds of mechanisms. At the same time, however, irreducible complexity also provides positive evidence for intelligent design.

Why irreducible complexity is a positive prediction for design. Irreducible complexity provides positive evidence for intelligent design because "it is a special case of specified complexity"[75] and "the defining feature of intelligent causes is their ability to create novel information and, in particular, specified complexity."[76] As Stephen Meyer elaborates, "whenever large amounts of specified complexity or information content are present in an artifact or entity whose causal story is known, invariably creative intelligence—design—has played a causal role in the origin of that entity."[77] Thus, irreducible complexity (i.e., specified complexity) provides positive, empirical evidence for design.

Scott Minnich explained this point at trial during cross-examination:

> In other words, you're saying, it's an argument out of ignorance. And I don't think it is... it's an argument out of our common cause and effect experience where we find these machines or information

---

74. Michael Behe, *Molecular Machines: Experimental Support for the Design Inference* at http://www.arn.org/docs/behe/mb_mm92496.htm. (last visited Jan. 30, 2006).

75. William A. Dembski, No Free Lunch 115 (Oxford, Rowman & Littlefield 2002).

76. *Id.* at xiv.

77. Stephen C. Meyer, *Science and Evidence for Design in Physics and Biology: From the Origin of the Universe to the Origin of Life, in* 9 The Proceedings of the Wethersfield Institute (San Francisco, Ignatius Press 2000).

storage systems. From our experience, we know there's an intelligence behind it.[78]

Thus Minnich characterized the inference to design as based upon our positive knowledge that intelligent agents habitually can and do produce machines (i.e., things which exhibit specified complexity). This is neither simply a negative argument against evolution nor a form of "contrived dualism." Unfortunately, Judge Jones misrepresented Minnich's real position in his opinion, claiming that Minnich "conceded" that "[i]rreducible complexity is a negative argument against evolution, not proof of design...."[79] Minnich did so such thing.

Had Judge Jones paid more attention to the arguments actually made by ID proponents, it would have been clear to him why irreducible complexity can be an argument for design as well as a criticism of Darwinian evolution. This is a genuine dualism due to the fact that neo-Darwinism and intelligent design make competing and opposite predictions about irreducible complexity.

3. "ID's negative attacks on evolution have been refuted by the scientific community."

Judge Jones shows no awareness that many of the "negative" scientific arguments made by ID proponents against the sufficiency of natural selection and random mutation to explain the complexity of life are also made by many evolutionists and other scientists who do not support intelligent design.[80] Over 500 doctoral scientists have signed a statement

---

78. Transcript of Testimony of Scott Minnich 86, *Kitzmiller*, No. 4:04-CV-2688 (M.D. Pa., Nov. 3, 2005).

79. Kitzmiller, 2005 WL 3465563, at *29.

80. For example, *see* signers of the Brief of Amicus Curiae Biologists and Other Georgia Scientists, In Support Of Defendants, 9-11, *Selman v. Cobb County School District* (No. 1:02-CV-2325-CC), *at* http://www.discovery.org/scripts/viewDB/filesDB-download.php?command=download&id=617. *See also* Brief of Amici Curiae Biologists And Other Scientists in support of Defendants at 1–2, *Kitzmiller* (No. 4:04-cv-2688), *at* http://www.discovery.org/scripts/viewDB/filesDB-download.php?command=download&id=558 (signed by

of "dissent" from Darwinism on precisely this point. (Some of the scientists also endorse ID, others do not.) The signers of the dissent from Darwin statement include distinguished scientists at such institutions as Princeton, MIT, the University of Georgia, the Ohio State University, the University of Wisconsin, and members of the U.S. and Russian national academies of science.[81]

"Some defenders of Darwinism embrace standards of evidence for evolution that as scientists they would never accept in other circumstances," said signer Henry Schaeffer, director of the Center for Computational Quantum Chemistry at the University of Georgia.[82] Other leading scientists go even further. "The ideology and philosophy of neo-Darwinism, which is sold by its adepts as a scientific theoretical foundation of biology, seriously hampers the development of science and hides from students the field's real problems," stated Vladimir L. Voeikov, Professor of Bio-organic Chemistry at Lomonosov Moscow State University.[83] It would be news to these scientists that critiques of Darwinian natural selection have been "refuted."

Even on the specific issue of irreducible complexity, which is a focus of Judge Jones' opinion, the evidence does not bear out Jones' broad claim. While some biologists have challenged Michael Behe's claims about the lack of Darwinian pathways and irreducible complexity, others have conceded them, and Behe himself has responded to his critics

---

some scientists who are skeptical of intelligent design who nonetheless feel it is appropriate to question Darwin or teach intelligent design in schools.)

81. *A Scientific Dissent from Darwinism, at* www.dissentfromdarwin.org/ (last visited Mar. 1, 2006).

82. 100 Scientists, National Poll Challenge Darwinism, *at* http://www.review-evolution.com/press/pressRelease_100Scientists.php (last visited Jan. 30, 2006).

83. *Over 500 Scientists Convinced by New Scientific Evidence That Darwinian Evolution is Deficient, at* http://www.discovery.org/scripts/viewDB/index.php?command=view&id=2732 (last visited Jan. 30, 2006)

in detail in peer-reviewed journals.[84] Shortly after Behe's *Darwin's Black Box* came out in 1996, biochemist James Shapiro of the University of Chicago acknowledged that "there are no detailed Darwinian accounts for the evolution of any fundamental biochemical or cellular system, only a variety of wishful speculations."[85] Five years later in a scientific monograph published by Oxford University Press, biochemist Franklin Harold, who rejects intelligent design, similarly admitted: "We must concede that there are presently no detailed Darwinian accounts of the evolution of any biochemical or cellular system, only a variety of wishful speculations."[86]

Other scientists have begun to cite Behe's ideas favorably in their own scientific publications. In a 2001 technical article in *Nature's* peer-reviewed *Encyclopedia of the Life Sciences*, two biologists cited Behe's identification of "irreducibly complex structures," as evidence for the possible limits of the mutation-selection mechanism, noting that "[u]p to now, none of these systems [described by Behe] has been satisfactorily explained by neo-Darwinism."[87] Similarly, a 2002 article published in the peer-reviewed *Annals of the New York Academy of Sciences* stated that while "Michael Behe may have been overly hasty in dismissing the possibility of the evolution of such mechanisms by natural selection… his notion of 'irreducible complexity' surely captured a feature of developmental systems that is of major importance."[88] What one has here is

84. See Michael J. Behe, *Reply to My Critics: A Response to Reviews of Darwin's Black Box: The Biochemical Challenge to Evolution*, 16 Biology and Philosophy 685–709 (2001); Michael J. Behe, *Self-Organization and Irreducibly Complex Systems: A Reply to Shanks and Joplin*, 67 Philosophy of Science 155-162 (March, 2000).

85. James A. Shapiro, "In the details . . . what?" National Review 62–65. (Sept. 16, 1996).

86. Franklin M. Harold, The Way of the Cell: Molecules, Organisms and the Order of Life 205 (New York, Oxford University Press 2001).

87. Heinz-Albert Becker & Wolf-Ekkehard Loennig, *Transposons: Eukaryotic*, Encyclopedia of Life Sciences 8 (Nature Publishing Group 2001).

88. Evelyn Fox Keller, *Developmental Robustness*, 981 Ann. N.Y. Acad. Sci. 189, 190 (2002) (emphasis added).

evidence of a scientific debate, a situation far different from the Judge's hamfisted and simply inaccurate assertion that Behe's arguments have been "refuted."

Indeed, to determine whether Behe has in fact been "refuted" one would have to carefully sift the evidence on both sides of the question, something Judge Jones was loathe to do. Instead, he was quick to canonize a speculative explanation offered by Darwinian biologist Kenneth Miller during the trial to explain away the irreducible complexity of the bacterial flagellum:

> [W]ith regard to the bacterial flagellum, Dr. Miller pointed to peer reviewed studies that identified a possible precursor to the bacterial flagellum, a subsystem that was fully functional, namely the Type-III Secretory System.[89]

As even Judge Jones concedes, however, there is no scientific consensus about whether the Type-III Secretory System is *in fact* a precursor of the bacterial flagellum. Given this situation, it is puzzling that Jones would cite such an admittedly speculative claim as evidence that Behe has been "refuted" by the scientific community. Jones further ignored testimony by microbiologist Scott Minnich, who explained that even if Miller's speculative scenario turned out to be true, it would not be sufficient to prove a Darwinian explanation for the origin of the flagellum, because there is still a huge leap in complexity from a Type-III Secretory System to a flagellum:

> Q. Would it be fair to say that if the type three secretory system was found to have preceded the bacterial flagellum, we'd still have difficulty with trying to determine how that one system that functions as a secretory system could then become a separate system that functions as a motor, flagellar motor?

---

89. *Kitzmiller,* 2005 WL 3465563, at *30.

A. Right... having a nano syringe and developing that into a rotary engine, you know, is a big leap.[90]

Minnich further explained in his testimony why so-called "co-optation" does not refute irreducible complexity, testimony which Jones also ignored.[91]

The unresolved challenge irreducible complexity continues to pose for Darwinian evolution is starkly summarized by mathematician William Dembski:

> [F]inding a subsystem of a functional system that performs some other function is hardly an argument for the original system evolving from that other system. One might just as well say that because the motor of a motorcycle can be used as a blender, therefore the [blender] motor evolved into the motorcycle. *Perhaps, but not without intelligent design.* Indeed, multipart, tightly integrated functional systems almost invariably contain multipart subsystems that serve some different function. At best the TTSS [Type-III Secretory System] represents one possible step in the indirect Darwinian evolution of the bacterial flagellum. But that still wouldn't constitute a solution to the evolution of the bacterial flagellum. What's needed is a complete evolutionary path and not merely a possible oasis along the way. To claim otherwise is like saying we can travel by foot from Los Angeles to Tokyo because we've discovered the Hawaiian Islands. Evolutionary biology needs to do better than that.[92]

It seems evident from Judge Jones' opinion that he reserved all of his skepticism for the proponents of intelligent design, while he was com-

---

90. Transcript of Testimony of Scott Minnich 112, *Kitzmiller*, No. 4:04-CV-2688 (M.D. Pa., Nov. 3, 2005).

91. *Id* at 102, *Kitzmiller*.

92. William A. Dembski, *Rebuttal to Reports by Opposing Expert Witnesses*, at http://www.designinference.com/documents/2005.09.Expert_Rebuttal_Dembski.pdf (last visited Jan. 30, 2006). This document was not offered at trial, but with regards to the TTSS it succinctly summarizes the arguments made by Minnich in his lengthy testimony.

pletely credulous when it came to any assertion made by the defenders of neo-Darwinism. Time and again he accepted at face value claims
made by evolutionists, while paying little or no attention to the evidence
and arguments submitted by scientists supporting ID.[93] Consider his
skewed summary of the evidence relating to the irreducible complexity
of the immune system.[94] He cited Kenneth Miller's speculative assertions as if they were facts, while refusing even to mention biochemist
Michael Behe's detailed rebuttal during the trial. As Behe points out:

> In my own direct testimony I went through the papers refer-
> enced by Professor Miller in his testimony and showed they didn't
> even contain the phrase "random mutation"; that is, they assumed
> Darwinian evolution by random mutation and natural selection was
> true—they did not even try to demonstrate it. I further showed in
> particular that several very recent immunology papers cited by Mill-
> er were highly speculative, in other words, that there is no current
> rigorous Darwinian explanation for the immune system. The Court
> does not mention this testimony.[95]

Judge Jones appears to have adopted a "hear no evil, see no evil" at-
titude to evidence challenging neo-Darwinism. His refusal to fairly consider both sides of the evidence in the case made his findings a foregone
conclusion.

4. "ID has failed to gain acceptance in the scientific community."

If "gain acceptance" means attracting the support of credible scientists at mainstream universities and the production of serious research

---

93. It bears repeating that in many cases a judge *must* determine the relative
persuasiveness of scientific testimony, and "choose sides" so as to resolve a liti-
gated claim. But Judge Jones clearly failed to grasp the distinction between his
role as a finder of disputed fact and his ambition to be the arbiter of what
constitutes science.

94. *Kitzmiller*, 2005 WL 3465563 at *31.

95. See appendix A.

and scholarship, then ID has in fact started to gain acceptance in the scientific community. ID supporters are publishing their work in mainstream academic and scientific publications (see Appendix B), and ID proponents include biologists, biochemists, physicists, astronomers, philosophers of science and other scholars at major research institutions such as UCLA, the University of Georgia, the University of Minnesota, and Iowa State University. There is also a professional society for scientists who are sympathetic to intelligent design, the International Society for Complexity, Information and Design (ISCID). ISCID publishes a technical journal and has nearly 60 Fellows, most of whom have their primary affiliations with major universities around the world.[96] Ironically, the Dover trial itself showcased how intelligent design has successfully drawn support from mainstream scientists. Michael Behe and Scott Minnich, the two expert witnesses who testified on behalf of ID at the trial, are both tenured biologists at reputable secular universities who have impeccable records of scientific research and publication.

Of course, if "acceptance" means attracting the support of the majority of scientists, especially Darwinian biologists, then ID obviously "has failed to gain acceptance." There is no question that intelligent design is a minority scientific view. But as noted earlier, all new scientific theories start off as a minority viewpoint, and disqualifying any theory from being "scientific" until it gains majority support among science would be preposterous.

Judge Jones' perspective on scientific debate seems simplistic and naive. He assumes that new scientific theories will be discussed without rancor or efforts to silence dissenters. Yet the history of science demonstrates the opposite. New ideas are often bitterly opposed by the champions of the existing orthodoxy in science, and mainstream science journals frequently refuse to publish ideas outside of the existing paradigm. As renowned historian of science Thomas Kuhn points out:

---

96. See *Society Fellows, at* http://www.iscid.org/fellows.php (last visited Jan. 30, 2006).

No part of the aim of normal science is to call forth new sorts of phenomena; indeed those that will not fit the box are often not seen at all. Nor do scientists normally aim to invent new theories, and they are often intolerant of those invented by others.[97]

In particular, scientists with "productive careers" may exhibit "[l]ifelong resistance" to new paradigms in science.[98] Scientists who established their careers as defenders of Darwinism can hardly be expected to give up their positions happily, especially if they have written textbooks on the topic. (Two rare exceptions are biophysicist Dean Kenyon of San Francisco State University and evolutionary biologist Stanley Salthe of Binghamton University, State University of New York.) Many federal and foundation research grants are pegged to Darwinism. Instead of welcoming criticism, the defenders of the dominant paradigm have circled the wagons in just the fashion that Kuhn's paradigm model anticipates.

Strident opposition to a new paradigm is precisely what biochemist Michael Behe and other pro-ID scientists have faced as they have sought to research and publish their ideas about intelligent design. An editor of a science journal specifically apologized to Behe for being unable to publish his article because it was not "orthodox:"

> I'm torn by your request to submit a (thoughtful) response to critics of your non-evolutionary theory for the origin of complexity. On the one hand I am painfully aware of the close-mindedness of the scientific community to non-orthodoxy, and I think it is counterproductive. But on the other hand we have fixed page limits for each month's issue, and there are many more good submissions than we can accept. So, your unorthodox theory would have to displace

---

97. THOMAS KUHN, THE STRUCTURE OF SCIENTIFIC REVOLUTIONS 24 (2nd ed. Chicago, University of Chicago Press 1970).

98. *Id.* at 141.

something that would be extending the current paradigm... You are in for some tough sledding. [99]

The cost of supporting a new paradigm in science can be much higher than letters of rejection from science journals.[100] As Judge Jones was made aware,[101] a growing number of scientists who are sympathetic or fair-minded to intelligent design are facing vilification, harassment, and even loss of employment.

Richard Sternberg is a trained evolutionary biologist,[102] with two doctorates, and a former editor of the peer-reviewed biology journal, *Proceedings of the Biological Society of Washington* (PBSW). As a PBSW editor, in 2004 Dr. Sternberg oversaw the publication of a peer-reviewed technical article which supported the hypothesis of intelligent design.[103] Although the article was reviewed and published using normal procedures,[104] Dr. Sternberg subsequently experienced retaliation by his co-workers and superiors at the Smithsonian, including transfer to a hostile supervisor, removal of his name placard from his door, deprivation of workspace, subjection to work requirements not imposed on others, re-

99. Michael J. Behe, *Correspondence With Science Journals: Response to Critics Concerning Peer Review, at* http://www.discovery.org/scripts/viewDB/index. php?command=view&id=450 (last visited Jan. 30, 2006) (quoting an unnamed editor of a mainstream scientific journal regarding why the editor could not publish Behe's paper challenging the evolution of the blood clotting cascade).

100. For an account of modern-day persecution of scientists, *see* Gordon Moran, Silencing Scientists and Scholars in Other Fields: Power, Paradigm Controls, Peer Review, and Scholarly Communication (Greenwich, Connecticut, Ablex Publishing Corporation 1998).

101. Brief of Amici Curiae Biologists And Other Scientists In support of Defendants at 23-28, *Kitzmiller* (No. 4:04-cv-2688), *at* http://www.discovery. org/scripts/viewDB/filesDB-download.php?command=download&id=558 (last visited Jan. 30, 2006). Included in this book as Appendix C.

102. Dr. Sternberg holds Ph.D.'s in molecular evolution and theoretical biology. *See* http://www.rsternberg.net/CV.htm (last visited Sept. 9, 2005).

103. Meyer, *supra* note 59.

104. *See* rsternberg.net/ (last visited Sept. 9, 2005). *See also* http://www.rsternberg.net/OSC_ltr.htm (last visited Sept. 9, 2005).

striction of specimen access, and loss of his keys.[105] Smithsonian officials also tried to smear Dr. Sternberg's reputation[106] and even investigated his religious and political affiliations in violation of his privacy and First Amendment rights.[107] According to an investigation by the U.S. Office of Special Counsel (OSC), these efforts were aimed at creating "a hostile work environment… with the ultimate goal of forcing [Sternberg]… out of the [Smithsonian]."[108] Furthermore, the OSC found that the pro-evolution National Center for Science Education (NCSE) helped devise the strategy to have Dr. Sternberg "investigated and discredited."[109] NCSE executive director Eugenie Scott later indicated to the *Washington Post* that Sternberg was lucky he was not fired outright: "If this was a corporation, and an employee did something that really embarrassed the administration… how long do you think that person would be employed?"[110]

Another target of intimidation has been Guillermo Gonzalez, an astronomer at Iowa State University (ISU). As the co-author of *The Privileged Planet*, Dr. Gonzalez postulated that the laws of the universe were intelligently designed to permit the existence of advanced forms of life.[111] Some of Dr. Gonzalez's astronomical work fundamental to

105. *Id.*

106. Michael Powell, *Editor Explains Reasons for 'Intelligent Design' Article*, WASHINGTON POST, August 19, 2005, A19. At http://www.washingtonpost.com/wp-dyn/content/article/2005/08/18/AR2005081801680_3.html (last visited Sept. 15, 2005).

107. *Id.*

108. See http://www.rsternberg.net/ (last visited Sept. 9, 2005). *See also* http://www.rsternberg.net/OSC_ltr.htm (last visited Sept. 9, 2005).

109. *Id.*

110. Michael Powell, *Editor Explains Reasons for 'Intelligent Design' Article*, WASHINGTON POST, August 19, 2005, A19. At http://www.washingtonpost.com/wp-dyn/content/article/2005/08/18/AR2005081801680_3.html (last visited Sept. 15, 2005).

111. GUILLERMO GONZALEZ AND JAY W. RICHARDS, THE PRIVILEGED PLANET: HOW OUR PLACE IN THE COSMOS IS DESIGNED FOR DISCOVERY (Washington, D.C., Regnery Publishing 2004).

his design hypotheses appeared as a cover story in *Scientific American*.[112] His book featured endorsements from leading scientists at Harvard and Cambridge, and was lauded in a review by David Hughes, a Vice-President of the Royal Astronomical Society.[113] Yet in retaliation against Dr. Gonzalez's application of design to astronomy, his opponents at ISU circulated a petition signed by over 120 faculty members "denouncing 'intelligent design.'"[114] The leader of the intimidation campaign—also faculty adviser for the ISU Atheist and Agnostic Society[115]—accused Gonzalez of having a hidden religious agenda. Others similarly "charged him with forcing his scientific evidence into a religious prism, fingering him as an academic fraud."[116] Like Sternberg, Gonzalez's attempts to focus on science have been futile: "I don't bring God into science. I've looked out at nature and discovered this pattern, based on empirical evidence."[117] After initiating the campaign of harassment, Gonzalez's chief accuser castigated Gonzalez for declining to appear at a lopsided "forum" sponsored by critics determined to denounce intelligent design.[118] Since he is coming up for tenure in the near future, Gonzalez is especially vulnerable to this effort to create a hostile work environment.

112. Guillermo Gonzalez et al., *Refuges for Life in a Hostile Universe*, Scientific American 60–67 (Oct., 2001).

113. See endorsements at http://www.privilegedplanet.com/endorsements.php (last visited Feb. 22, 2006). David Hughes' review appeared in The Observatory 113 (April 2005).

114. Jamie Schuman, *120 Professors at Iowa State U. Sign Statement Criticizing Intelligent-Design Theory*, Chronicle of Higher Education, August 26, 2005. At http://www.chronicle.com/temp/email.php?id=7d6oum55u2gs4xg z0zoqckkx4ulkgoy6 (last visited Sept. 9, 2005).

115. *Id.*

116. Reid Forgrave, *Life: A universal debate*, Des Moines Register, August 31, 2005, at http://www.dmregister.com/apps/pbcs.dll/article?AID=/20050831/ LIFE/%20508310325/1001/LIFE (last visited Sept. 12, 2005).

117. *Id.*

118. Lisa Livermore, *'Intelligent design' faces ISU opposition*, Des Moines Register, August 26, 2005, at http://www.desmoinesregister.com/apps/pbcs. dll/article?AID=/20050826/NEWS02/508260394/1001 (last visited Sept. 9, 2005).

Other scientists have experienced even worse retribution. Dr. Caroline Crocker was a biology professor at George Mason University until she favorably mentioned intelligent design in a class and was then banned from teaching both intelligent design *and* evolution.[119] Subsequently, her contract was not renewed, and she lost her job. Leading design theorist Dr. William Dembski was banned from teaching at Baylor University and forced into a "five-year sabbatical."[120] This followed a letter-writing campaign by Barbara Forrest (whose testimony in Dover Judge Jones hailed) to dissuade scholars from associating with Dembski's Polanyi Center at Baylor, because she alleged it was "the most recent offspring of the creationist movement."[121] In another case, Dr. Nancy Bryson was removed as head of the Division of Science and Mathematics at Mississippi University for Women, without explanation, the day after she taught an honors forum entitled "Critical Thinking on Evolution."[122]

ID is clearly experiencing an extreme form of the kind of opposition every new scientific theory experiences when it challenges a deeply-entrenched paradigm. As noted during the Dover trial, the "Big Bang" theory about the origins of the universe initially faced similar objections to the ones being lodged against ID.[123] The Big Bang was bitterly resisted by much of the scientific community for several decades, largely due to its presumed philosophical and theological implications. As famed astronomer Arthur Eddington wrote in the prestigious science journal *Nature*, "[p]hilosophically, the notion of a beginning of the present order

---

119. Geoff Brumfiel, *Cast out from class*, 434 NATURE 1064 (April 28, 2005).

120. *Id.*

121. Barbara Forrest, *Letter to Simon Blackburn*, at http://www.designinference. com/documents/2005.05.ID_at_Baylor.htm (last visited Sept. 9, 2005).

122. Transcript of Proceedings before Kansas State Board of Education 3-9, *at* http://www.ksde.org/outcomes/schearing05072005am.pdf (last visited Sept. 15, 2005).

123. Transcript of Testimony of Michael Behe 117-123, *Kitzmiller*, No. 4:04-CV-2688 (M.D. Pa., Oct. 17, 2005).

of Nature [as implied by the Big Bang] is repugnant to me... I should like to find a genuine loophole."[124]

It is fortunate for science that no federal judge took it upon himself at the time to declare whether the Big Bang theory qualified as science or was simply "creationism in disguise."

5. "ID... has not generated peer-reviewed publications."[125]

Judge Jones writes that "a final indicator of how ID has failed is the *complete absence* of peer-reviewed publications supporting the theory."[126] Again, he confidently asserts that "ID is not supported by *any* peer-reviewed research, data or publications."[127] In a footnote, he glancingly mentions one peer-reviewed article in the journal *Protein Science* by Michael Behe, but complains that this article does not explicitly reference ID.[128]

Judge Jones shows no awareness of several other peer-reviewed publications explicitly supporting both intelligent design and Behe's idea of irreducible complexity, even though a list of these publications was submitted to him as part of the official record in the case.[129] This list included such articles as Stephen Meyer's peer-reviewed technical article on the Cambrian explosion and intelligent design in *The Proceedings of the Biological Society of Washington*, a peer-reviewed biology journal, and a more recent technical article on irreducible complexity and intelligent

---

124. Sir Arthur Eddington, 127 NATURE 450 (1931).

125. Kitzmiller, 2005 WL 3465563, at *26. Judge Jones also writes "it has failed to publish in peer-reviewed journals." Id. at *34.

126. *Kitzmiller*, 2005 WL 3465563, at *34 (emphasis added).

127. *Id.* (emphasis added).

128. *Id.* n. 17.

129. Brief of Amicus Curiae FTE appendix D at 8–18, *Kitzmiller* (No. 4:04-cv-2688), at http://www.discovery.org/scripts/viewDB/filesDB-download.php?command=download&id=649 (last visited Jan. 30, 2006).

design in *Dynamical Genetics*.[130] Judge Jones did not deny that these articles were peer-reviewed. He simply ignored them. He ignored in addition peer-reviewed academic books like William Dembski's *The Design Inference* (Cambridge University Press) and Campbell and Meyer's *Darwinism, Design and Public Education* (Michigan State University Press).

At the trial, expert witness Scott Minnich testified that there were between "seven and ten" peer-reviewed papers supporting ID.[131] Jones chose to ignore this testimony as well. Jones implies by his wording that *either* peer-reviewed "publications" *or* "data" would count as legitimate scientific support for intelligent design. Yet both Behe and Minnich also testified about still other papers which provided data supporting ID, even if they did not use the word "intelligent design."[132] It would require only one of these articles to refute Judge Jones' assertion that intelligent design has produced no peer-reviewed publications. Judge Jones selected the facts as he pleased by ignoring the facts presented to him at trial and in amicus briefs. (For a list of peer-reviewed publications supporting intelligent design, see Appendix B.)

Thus far we have assumed for the sake of argument that Judge Jones was correct in relying on peer-review as a criterion of whether a theory is "scientific." But is his standard really valid? According to Judge Jones, "[e]xpert testimony revealed that the peer review process is 'exquisitely important' in the scientific process."[133] He then noted that "[p]eer review helps to ensure that research papers are scientifically accurate, meet the standards of the scientific method, and are relevant to other scientists

---

130. Lönnig, W.-E. *Dynamic genomes, morphological stasis and the origin of irreducible complexity,* in DYNAMICAL GENETICS 101-119 (Valerio Parisi et al. eds. 2004).

131. Transcript of Testimony of Scott Minnich, 34, *Kitzmiller*, No. 4:04-CV-2688 (M.D. Pa., Nov. 4, 2005).

132. Transcript of Testimony of Michael Behe 62, *Kitzmiller*, No. 4:04-CV-2688 (M.D. Pa., Oct. 18, 2005); Transcript of Testimony of Scott Minnich 34, 85, Kitzmiller, No. 4:04-CV-2688 (M.D. Pa., Nov. 4, 2005).

133. *Kitzmiller*, 2005 WL 3465563, at *34.

in the field." More generally, peer review helps determine "whether the researcher has employed sound science."

While these points are surely true, Judge Jones shows no knowledge of the recent origins of the peer-review process, nor of its significant weaknesses. Frank Tipler, a professor of mathematical physics at Tulane University, has noted that the system of peer-review for scientific publications so extolled by Judge Jones developed for the most part after World War II.[134] Does this mean that no scientific theory proposed prior to the late 1940s should qualify as genuine science? To ask such a question is to expose its patent absurdity. Tipler adds that "in the last several decades, many outstanding scientists have complained that their best ideas—the very ideas that brought them fame—were rejected by refereed journals," and laments that in many cases the so-called peer-reviewer "is… not as intellectually able as the author whose work he judges. We have pygmies standing in judgment on giants."[135] We also have peer-reviewers with predictable prejudices.

Judge Jones shows no awareness that the U.S. Supreme Court has articulated a different perspective on peer review than the one he advances. In *Daubert v. Merrel Dow Pharmaceuticals*, the U.S. Supreme Court held that "[p]ublication… is not a *sine qua non* of admissibility; it does not necessarily correlate with reliability."[136] According to the Supreme Court, peer-reviewed publication is a "relevant, though not dispositive" consideration of whether or not some scientific finding will be admitted into evidence.[137] The simple basis for this conclusion is that peer review is not necessarily the only indicator of reliability, because "in some instances well-grounded but innovative theories will not have been

---

134. Frank J. Tipler, *Refereed Journals: Do They Insure Quality or Enforce Orthodoxy?*, in Uncommon Dissent: Intellectuals Who find Darwinism Unconvincing 116 (William Dembski, ed. (Wilmington, Intercollegiate Studies Institute 2004).

135. *Id.*

136. *Daubert*, 509 U.S. at 593.

137. *Id.* at 594.

published."[138] An *amicus curiae* brief submitted by various scientists, including the late Stephen Jay Gould, captured the argument to allow non-mainstream views into the courtroom:

> Judgments based on scientific evidence, whether made in a laboratory or a courtroom, are undermined by a categorical refusal even to consider research or views that contradict someone's notion of the prevailing "consensus" of scientific opinion.... Automatically rejecting dissenting views that challenge the conventional wisdom is a dangerous fallacy, for almost every generally accepted view was once deemed eccentric or heretical. Perpetuating the reign of a supposed scientific orthodoxy in this way, whether in a research laboratory or in a courtroom, is profoundly inimical to the search for truth. A categorical refusal even to examine and consider scientific evidence that conflicts with some ill-defined notion of majority opinion is a recipe for error in any forum.... The quality of a scientific approach or opinion depends on the strength of its factual premises and on the depth and consistency of its reasoning, not on its appearance in a particular journal or on its popularity among other scientists.[139]

In fact, the Supreme Court in *Daubert v. Merrell Dow Pharmaceuticals* overruled the Ninth Circuit precisely because it "had refused to admit reanalyses of epidemiological studies that had been neither published nor subjected to peer review."[140] The Ninth Circuit excluded these studies because they were "unpublished, not subjected to the normal peer review process and generated solely for use in litigation."[141] According to the Supreme Court and various prominent scientists, it seems

138. *Id.* at 593.

139. Brief *Amici Curiae* of Ronald Bayer, Stephen Jay Gould, Gerald Holton, Peter Infante, Philip Landrigan, Everett Mendelsohn, Robert Morris, Herbert Needleman, Dorothy Nelkin, William Nicholson, Kathleen Joy Propert, and David Rosner, in support of petitioners, *Daubert*, 509 U.S. 579 (1993) (No. 92-102).

140. *Daubert*, 509 U.S. at 584.

141. *Daubert v. Merrell Dow Pharmaceuticals*, 951 F.2d 951 1128, 1131 (9th Circuit, 1991), overruled 509 U.S. 579 (1993).

clear that something *can* be scientific even if it has not been published in peer-reviewed journals.

6. "ID [has not]… been the subject of testing and research."

Testability is an important aspect of any scientific theory. Unless a theory can be tested, it is difficult if not impossible for an observer to determine whether it is true. However, testability is difficult to measure in an absolute way. Testability in science is correctly viewed in the "abstract" sense of whether a claim is truly testable *in principle*. As the eminent philosopher of science Phillip Quinn notes, "[t]he requirement is that a scientific theory be testable, not that its proponents actually test it."[142]

Even if the criterion were whether or not the theory of intelligent design has actually been tested, Judge Jones was in error. As noted previously, irreducible complexity is an argument for design, and Scott Minnich addressed the question of whether he has tested the irreducible complexity of the bacterial flagellum. He has. He testified and showed data of his own mutagenesis experiments which have determined that with respect to its full complement of genes, the bacterial flagellum is irreducibly complex.[143] Moreover, the claims of design proponents are eminently testable: one can easily test for irreducible complexity using knockout experiments, or specified complexity using mutational sensitivity tests. This can be done in principle, and has been done in practice. Minnich even mentioned the mutational sensitivity tests performed by biochemist Douglas Axe.[144]

Kenneth Miller tried to sidestep this obvious point in his expert testimony at the Dover trial by conceding that Behe's flagellum argument

---

142. P. L. Quinn, "The Philosopher of science as expert witness," in SCIENCE AND REALITY: WORKS IN THE PHILOSOPHY OF SCIENCE 32-53, (J. Cushing et al. eds., University of Notre Dame Press 1984).

143. Transcript of Testimony of Scott Minnich 103–112, *Kitzmiller*, No. 4:04-CV-2688 (M.D. Pa., Nov. 3, 2005).

144. Transcript of Testimony of Scott Minnich 34, *Kitzmiller*, No. 4:04-CV-2688 (M.D. Pa., Nov. 4, 2005).

was testable but insisting that it was a purely negative argument against Neo-Darwinism, not a positive case for intelligent design. This is mere wishful thinking on Miller's part. Behe's argument is also based on positive evidence for design. As noted earlier, Behe, Meyer, and other design theorists point to strongly positive grounds for inferring design from the presence of irreducibly complex machines and circuits in the non-living realm. This testable evidence is so powerful, so nearly ubiquitous, that it is often overlooked. Go out and find irreducibly complex machines, then seek out their causal history. Again and again, where their history is available to us (such as with the rotary engines made by the Mazda Corporation) one will find that they were designed by intelligent agents. Indeed, every time we know the causal history of an irreducibly complex system, it always turns out to have been the product of an intelligent cause. Both Behe and Minnich discussed this in their expert testimony, but Judge Jones apparently wasn't listening.

Summary. In the variety of ways that have been detailed above, Judge Jones failed to make a fair, impartial review of the evidence. Far from supporting his categorical conclusion that ID is not science, his analysis amounts to little more than an impassioned closing argument from Darwin's public defender.

CHAPTER III

KITZMILLER'S FAILURE TO TREAT RELIGION IN A NEUTRAL MANNER

Judge Jones based his ruling on the requirements of the Establishment Clause of the First Amendment, but he failed to observe the cardinal principle of the Establishment Clause, which is that religion must be treated in a *neutral* manner: "The First Amendment does not select any one group or any one type of religion for preferred treatment. It puts them all in [the same]... position."[145]

Judge Jones seemed to think that the possible religious implications of intelligent design theory made it a religious theory. He reached that conclusion apparently without even considering whether the religious implications of Darwinian evolution would yield the same conclusion. Similarly, he looked to the supposed religious motivations of the proponents of intelligent design theory to establish the religious nature of intelligent design theory without subjecting the proponents of Darwinian evolution to the same test. For example, many pages of Judge Jones' opinion are devoted to establishing the history of the "intelligent design movement" and the theological views of its advocates. He relies extensively on the testimony of Barbara Forrest, who "thoroughly and exhaustively chronicled the history of ID in her book and other writings for her testimony in this case."[146] There was no attempt to verify that

---

145. *U.S. v. Ballard*, 322 U.S. 78, 87 (1944).

146. *Kitzmiller*, 2005 WL 3465563, at *12–*13.

purported history and nowhere does Judge Jones subject Barbara Forrest to an examination of whether *her* background or *her* beliefs might be
relevant to the case. If Judge Jones wanted to play the motivation game,
he ought in fairness to have addressed the extensive evidence in one of
the amicus briefs documenting the anti-religious affiliations and motivations of many leading Darwinists, including especially Professor Forrest
herself.[147]

A. One-Sided (Non-Neutral) Treatment of Religious Implications

Design proponents have never pretended that intelligent design
lacked religious implications for many people. But Judge Jones failed to
address this issue in a neutral fashion, and he failed to recognize that
Darwinian evolution, just like intelligent design theory, has important
philosophical and even theological implications. Indeed, for every one
of the quotes Judge Jones relied upon to establish the religious nature of
intelligent design theory, one can find several statements by proponents
of Darwinism that argue for a parallel (but opposite) metaphysical conclusion.[148]

Oxford biologist Richard Dawkins, perhaps the world's most famous extant evolutionist, has famously stated, "Darwin made it pos-

---

147. Brief of Amici Curiae Biologists And Other Scientists in support of Defendants, at 20-23, *Kitzmiller* (No. 4:04-cv-2688), *at* http://www.discovery.
org/scripts/viewDB/filesDB-download.php?command=download&id=558
(last visited Jan. 30, 2006).

148. At the same time, it is important to stress that proponents of ID such as
the Discovery Institute have never argued that the theological or philosophical
implications of Darwinian theory should render it unsuitable for instruction
in the public schools. Quite the contrary, Discovery Institute has urged school
districts to teach public school students more about Darwinian evolution than
they do at present. See Discovery Institute's "Science Education Policy," at
http://www.discovery.org/scripts/viewDB/index.php?command=view&id=3
164&program=CSC%20-%20Science%20and%20Education%20Policy%20-
%20School%20District%20Policy (last visited Jan. 27, 2006).

sible to become an intellectually fulfilled atheist,"[149] while evolutionary biologist Douglas Futuyma has declared that "[b]y coupling undirected, purposeless variation to the blind, uncaring process of natural selection, Darwin made theological or spiritual explanations of the life processes superfluous."[150] And the writings of the late paleontologist Stephen Jay Gould are replete with such statements as:

+ [B]iology took away our status as paragons created in the image of God....[151]

+ Before Darwin, we thought that a benevolent God had created us.[152]

+ [T]he stumbling block to [accepting Darwin's theory] does not lie in any scientific difficulty, but rather in the radical philosophical content of Darwin's message... First, Darwin argues that evolution has no purpose... Second, Darwin maintained that evolution has no direction... Third, Darwin applied a consistent philosophy of materialism to his interpretation of nature. Matter is the ground of all existence; mind, spirit, and God as well, are just words that express the wondrous results of neuronal complexity.[153]

Cornell University evolutionist William Provine has similarly stated that, "Evolution is the greatest engine of atheism ever invented."[154]

Amazingly, the plaintiffs' own expert biologist Kenneth Miller drew a direct connection between evolution and philosophical materialism in the first two editions of one of his biology textbooks, in which he claimed:

---

149. RICHARD DAWKINS, THE BLIND WATCHMAKER 6 (New York, W. W. Norton, 1991).

150. DOUGLAS FUTUYMA, EVOLUTIONARY BIOLOGY 5 (3d ed., Sunderland, Sinauer Associates 1998).

151. STEPHEN JAY GOULD, EVER SINCE DARWIN 147 (New York, W. W. Norton 1977).

152. *Id* at 267.

153. *Id*. at 12–13.

154. William B. Provine, *Abstract of Will Provine's 1998 Darwin Day Keynote Address, Evolution: Free will and punishment and meaning in life.*

Darwin knew that accepting his theory required believing in *philosophical materialism,* the conviction that matter is the stuff of all existence and that all mental and spiritual phenomena are its by-products... Suddenly, humanity was reduced to just one more species in a world that cared nothing for us. The great human mind was no more than a mass of evolving neurons. Worst of all, there was no divine plan to guide us.[155]

Some of the above quotations were brought to Judge Jones' attention in the amicus briefs,[156] but Judge Jones was apparently interested only in the religious implications that flow from ID. Yet true neutrality requires that if a judge takes into account the philosophical or theological implications of intelligent design, he must also take into account the philosophical or theological implications of neo-Darwinism.

It is important to stress that our purpose here is *not* to suggest that Darwinists should be criticized by judges for pursuing an anti-religious agenda, but rather to show that a chief complaint against ID—that its religious implications disqualify it as science—is unsustainable because it would also disqualify Darwinian evolution as science.

B. One-Sided (Non-Neutral) Treatment of Secondary Effects

Judge Jones also used his one-sided analysis to selectively apply the principles of Establishment Clause jurisprudence concerning the primary and secondary effects of state action. The Supreme Court consistently

---

155. Joseph S. Levine and Kenneth R. Miller, Biology: Discovering Life 152 (1st ed., Lexington, D.C. Heath and Co. 1992). This language was not removed for the 2nd ed. in 1994, p. 161.

156. Brief of Amicus Curiae Discovery Institute at 28, *Kitzmiller* (No. 4:04-cv-2688), *at* http://www.discovery.org/scripts/viewDB/filesDB-download.php?command=download&id=558 (last visited Jan. 30, 2006) and Brief of Amicus Curiae FTE at 18, Kitzmiller (No. 4:04-cv-2688), *at* http://www.discovery.org/scripts/viewDB/filesDB-download.php?command=download&id=648 (last visited Jan. 30, 2006).

has held that the *primary* or direct effect of state action must be distinguished from incidental or secondary effects:

> The Court has made it abundantly clear, however, that "not every law that confers an 'indirect,' 'remote,' or 'incidental' benefit upon [religion] is, for that reason alone, constitutionally invalid." Here, whatever benefit there is to one faith or religion or to all religions, is indirect, remote, and incidental.[157]

State action that results in an indirect or secondary benefit (or harm) to religion is not unconstitutional according to the Supreme Court. In *Agostini v. Felton*,[158] the Court added that it is not the magnitude of the benefit that matters; the question is whether the effects/benefits of a policy provided are direct, or merely a consequence of implementing a religiously neutral or secular principle. If the latter, then the effect or benefit is incidental. As the Court held, a benefit to religion is merely incidental if it "is allocated on the basis of neutral, secular criteria that neither favor nor disfavor religion."[159]

Precisely such logic has permitted the courts at once to acknowledge the anti-religious implications of teaching neo-Darwinism[160] and at the same time to sanction its presentation in such unambiguous terms.[161] In the Dover case, plaintiffs freely admitted that the teaching of Darwinian evolution is offensive to certain religious beliefs[162] and their assertion of religious motivation is based on the claim that the conflict between neo-Darwinism and certain religious beliefs generated the Dover

---

157. *Lynch v. Donnelly*, 465 U.S. 668, 683 (1984) (internal citations omitted).

158. *Agostini v. Felton*, 521 U.S. 203, 231 (1997).

159. *Id.* (because services to students in a religious school resulted in a benefit that had been distributed on a neutral, secular basis, the program was constitutional).

160. *Epperson v. Arkansas*, 393 U.S. 97, 113 (1968) (Black, J., concurring).

161. *Id.*

162. Plaintiffs' Opposition to Motion for Summary Judgment at 59, *Kitzmiller et al. v. Dover Area School Board* (M.D. Pa., Dec. 20, 2005) (No. 4:04-CV-2688).

policy. More generally, as discussed, many neo-Darwinists have openly acknowledged the anti-theistic implications of their theory. Such statements raise an obvious question. As Justice Black asked in *Epperson v. Arkansas*, "[I]f the theory [of evolution] is considered anti-religious, as the Court indicates, how can the State be bound by the Federal Constitution to permit its teachers to advocate such an 'antireligious' doctrine to schoolchildren?"[163]

The answer to this rhetorical question is clear: The courts have treated the religious implications of neo-Darwinism as merely an incidental effect of the secular purpose of teaching students about a scientific theory.

But why shouldn't the courts treat intelligent design in the same manner? Despite any religious implications intelligent design may have, there are several neutral, secular criteria that could be the basis for including the theory of intelligent design in the science curriculum. For example, a school board might wish to (a) promote scientific literacy and (b) follow the Report language in the No Child Left Behind Act,[164] by including "the full range of scientific views" about biological evolution in its science curriculum. Even if one result of such a policy was to encourage (or discourage) various religious or philosophical beliefs, such effects, by the standard enunciated in *Agostini*, would be "merely incidental." Further, the variety of secular purposes for teaching about intelligent design cited in the previous section could generate other "neutral, secular criteria" that would justify teaching about intelligent design and render the effect of such a policy on religion merely incidental.

Yet in his ruling, Judge Jones arbitrarily treated the religious implications of intelligent design as if they were primary effects, not secondary

---

163. *Epperson*, 393 U.S. at 113 (Black, J., concurring).

164. "Where topics are taught that may generate controversy (such as biological evolution), the curriculum should help students to understand the full range of scientific views that exist..." Conference report to the No Child Left Behind Act, Congress; House Committee of Conference, *Report to Accompany H.R. 1*, 107th Congress, 1st sess., 78 (2001) H. Rept. 334, 78.

or incidental effects. Chapter 2 of this response explained why ID makes its claims based upon empirical data and scientific reasoning. Thus intelligent design should have been treated like Darwinian evolution, as a theory that should be assessed on its own terms, ones involving empirical evidence and appeals to the cause-and-effect structure of the world. If Jones looks at such theories by treating their religious implications as primary effects, then his rule would jeopardize the constitutionality of teaching any scientific theory, such as Neo-Darwinism or the Big Bang. This discriminatory double-standard cannot ultimately stand as a legal rule for teaching science in schools.

Further, no one "religion" is even implied by ID. Some Catholics, evangelical Christians, mainline Christians, Jews, Buddhists, and Muslims support intelligent design arguments, while others in every one of those categories do not. So where is the establishment of religion in the implications of the theory? Given that such pagan philosophers as Plato were advocates of design, it is hard to see that religion itself is necessarily implied by ID, much less some particular religion. Some creationists have leveled this very charge against intelligent design.

C. One-Sided (Non-Neutral) Treatment of Religious Motives

Let us now turn to the subject of motives. Despite spending pages and pages reviewing the supposed religious backgrounds and motivations of the "Intelligent Design Movement," Judge Jones devotes not a word to the religious (or anti-religious) motivations of the various advocates of Darwinian evolution. This is another example of the Judge's stunning double standard. He accepted plaintiffs' arguments that a scientist's religious beliefs could turn his scientific efforts into "mere religion," without applying the same test to those who support Darwinian evolution.

For example, Eugenie Scott, director of a leading activist organization opposing the teaching of intelligent design, the National Center for Science Education (NCSE), is a "Notable Signer" of the Humanist Manifesto III. The Manifesto makes broad theological (or "anti-theological") claims that "[h]umans are... the result of unguided evolutionary change. Humanists recognize nature as self-existing."[165]

NCSE official (and Dover expert witness) Barbara Forrest serves on the Board of Directors of the New Orleans Secular Humanist Association (NOSHA), which describes itself as "an affiliate of American Atheists, and [a] member of the Atheist Alliance International."[166] NOSHA is also an affiliate of the Council for Secular Humanism which it describes as "North America's leading organization for non-religious people."[167] NOSHA's links page boasts "The Secular Web," whose "mission is to defend and promote metaphysical naturalism, the view that our natural world is all that there is, a closed system in no need of an explanation and sufficient unto itself."[168] Most notably, NOSHA is an associate member of the American Humanist Association,[169] which publishes the Humanist Manifesto III.[170] In 1996, this American Humanist Association named Richard Dawkins as its "Humanist of the Year."[171] To help underscore the anti-religious mindset of these organizations, in his acceptance speech for the award before the American Humanist Associ-

---

165. Humanist Manifesto III Public Signers, *at* http://www.americanhumanist.org/3/HMsigners.htm (last visited Sept. 10, 2005) and Humanism and its Aspirations, *at* http://www.americanhumanist.org/3/HumandItsAspirations.htm (last visited Sept. 10, 2005).

166. New Orleans Secular Humanist Association home page, *at* http://nosha.secularhumanism.net/index.html (last visited Sept. 10, 2005). Forrest is listed as a member of the board of directors on the "Who's Who" page of the website, see http://nosha.secularhumanism.net/whoswho.html (last visited Sept. 10, 2005).

167. *Id.*

168. *Id.*

169. *Id.*

170. *See* http://www.americanhumanist.org/3/HumandItsAspirations.htm (last visited Sept. 10, 2005).

171. *See* http://www.thehumanist.org/humanist/articles/dawkins.html (last visited Sept. 10, 2005).

ation, Dawkins stated "faith is one of the world's great evils, comparable to the smallpox virus but harder to eradicate."[172]

Another public opponent of intelligent design is Nobel Laureate Steven Weinberg.[173] Weinberg explicitly says that his scientific career is motivated by a desire to disprove religion:

> I personally feel that the teaching of modern science is corrosive of religious belief, and I'm all for that! One of the things that in fact has driven me in my life, is the feeling that this is one of the great social functions of science—to free people from superstition.[174]

Lest there be any doubt about what Weinberg means by "superstition," he goes on to say that he hopes "that this progression of priests and ministers and rabbis and ulamas and imams and bonzes and bodhisattvas will come to an end, that we'll see no more of them. I hope that this is something to which science can contribute and if it is, then I think it may be the most important contribution that we can make."[175]

Even the eminent National Academy of Sciences, which has issued various booklets against teaching intelligent design,[176] has a member-

---

172. *Id.*

173. In 2003, Dr. Weinberg testified in support of teaching only the evidence for evolution before the Texas State Board of Education. *See* Forrest Wilder, "Academics need to get more involved," Opinion, *The Daily Texan*, Oct. 2, 2003, *at* http://www.dailytexanonline.com/media/paper410/news/2003/10/02/Opinion/Academics.Need.To.Get.More.Involved-510574.shtml (last visited Sept. 10, 2005).

174. Free People from Superstition, *at* http://www.ffrf.org/fttoday/2000/april2000/weinberg.html (last visited Sept. 15, 2005).

175. *Id.*

176 *See* NATIONAL ACADEMY OF SCIENCES, *supra* note 72; NATIONAL ACADEMY OF SCIENCES, SCIENCE AND CREATIONISM: A VIEW FROM THE NATIONAL ACADEMY OF SCIENCES (2nd ed., Washington, D.C., National Academy Press, 1999).

ship of biologists of whom nearly 95 percent describe themselves as atheists or agnostics.[177]

These anti-religious motivations are cited here not because they disqualify anyone from making a scientific argument, but to demonstrate that the personal beliefs of theists should similarly be ignored in determining whether their scientific claims will be given a fair hearing. Our contention is that religious or philosophical motivations, however strongly held or expressed, should have no legal significance in determining the scientific standing of a theory.

Judge Jones could attach legal significance to the evidence of religious motivation only by ignoring the evidence that was offered with respect to the anti-religious statements and affiliations of many advocates of Darwinism. In the *Kitzmiller* opinion religious neutrality was displaced by a sorry display of *ad hominem* attacks.

D. An Effort to Dictate a Particular Theological View of Evolution

Judge Jones also took sides in a theological dispute in a way that on its face contradicts an ostensible central theme of his opinion. Near the end of his ruling, he declared:

> Both Defendants and many of the leading proponents of ID make a bedrock assumption which is utterly false. Their presupposition is that evolutionary theory is antithetical to a belief in the existence of a supreme being and to religion in general.[178]

It is important to note that Judge Jones is flatly contradicting what he said earlier about the nature of the "intelligent design movement" and religion. As noted, much of his opinion is devoted to showing that opposition to evolutionary theory is motivated by religion. Yet on page 136 he says that Darwinian evolution and religion do not conflict. One can only

177. Edward J. Larson and Larry Witham, "Scientists and Religion in America," 281 Scientific American 88-93, (Sept., 1999).

178. *Kitzmiller*, 2005 WL 3465563, at *51.

make sense of this by interpreting Judge Jones to be saying that there is no conflict between evolution and *true religion*. In his own words, the view "that evolutionary theory is antithetical to a belief in the existence of a supreme being" is "*utterly false.*"[179] In essence, Judge Jones is declaring that to see conflict between religion and evolution is heresy. He seems to have forgotten that it is not the place of the law to declare what is orthodox and what is heretical.[180]

It is important to be very clear: the fact that *some forms* of evolution may conflict with *some forms* of religious belief does not require the public schools to refrain from presenting information that creates conflict with those religious beliefs. But precisely because of that potential conflict the state must act *neutrally*; it is no part of neutrality for a Judge to assert as a matter of legal fact that there is no conflict between evolution and religion, any more than it is the place of a judge to reassure the public that there is no conflict between serving God and serving in the U.S. military, or between the Christian religion and same-sex marriage.

---

179. *Id.* (emphasis added).

180. Under the principle of religious neutrality, courts are forbidden from passing judgment upon the validity of religious beliefs:

"The law knows no heresy, and is committed to the support of no dogma, the establishment of no sect." ... Freedom of thought, which includes freedom of religious belief, is basic in a society of free men. It embraces the right to maintain theories of life and of death and of the hereafter which are rank heresy to followers of the orthodox faiths. Heresy trials are foreign to our Constitution. Men may believe what they cannot prove. They may not be put to the proof of their religious doctrines or beliefs. "Religious experiences which are as real as life to some may be incomprehensible to others. Yet the fact that they may be beyond the ken of mortals does not mean that they can be made suspect before the law. ... The religious views espoused by respondents might seem incredible, if not preposterous, to most people. But if those doctrines are subject to trial before a jury charged with finding their truth or falsity, then the same can be done with the religious beliefs of any sect. When the triers of fact undertake that task, they enter a forbidden domain. The First Amendment does not select any one group or any one type of religion for preferred treatment. It puts them all in that position." *Ballard*, 322 U.S. at 86–7 (internal citations omitted).

It is a cruel irony that after spending more than a hundred pages inveighing against the supposed endorsement of religion by an inconsequential four-paragraph statement read to students, Judge Jones did not hesitate to declare—as a matter of federal law—that there is only one correct answer to a hotly disputed theological question. The fact that he buttressed his pronouncement with the expert testimony of the plaintiffs' theologian only compounds the absurdity of his ruling.

By far the greatest affront to the Establishment Clause in the Dover case was not any action of the Dover Area School Board, but Judge Jones' official endorsement (on behalf of the federal government's judicial branch) of a particular religious understanding of evolution.

In fairness, the double-standard exhibited by Judge Jones on religion is rampant among the defenders of evolution. For example, the National Center for Science Education, whose experts were on the front-line in the *Kitzmiller* case, helped develop an educational website that encourages teachers to use theological statements by religious groups endorsing evolution in order to convince students what their religious views about evolution should be.[181] Funded by more than a half-million dollars in federal tax dollars, the website is now the subject of a federal lawsuit charging that the creators of the website have violated the Establishment Clause by having the government endorse a particular religious view of evolution.[182] Unlike the lawsuit in Dover, this lawsuit against evolutionists has received virtually no coverage from the national newsmedia. Apparently journalists are not concerned about the government promoting religion in science class so long as religion is used to endorse evolution.

---

181. John G. West, *Evolving Double Standards*, National Review Online, April 1, 2004, *at* http://www.discovery.org/scripts/viewDB/index. php?command=view&id=1967 (last visited Jan. 30, 2006) and Francis J. Beckwith, *Government-Sponsored Theology*, American Spectator, April 7, 2005, at http://www.discovery.org/scripts/viewDB/index. php?command=view&id=1990 (last visited Jan. 30, 2006).

182. Complaint at 4, *Caldwell v. Caldwell et al.* (N.D. Cal., Oct. 13, 2005) (No. C05-04166 PJH). *See also Quality Science Education for All in the News, at* http://qsea.org/_wsn/page5.html (last visited Jan. 30, 2006).

This is a great situation for Darwinists. On the one hand, evolutionists like Professor Provine declare that evolution is a tool for inculcating atheism, while on the other hand, the NCSE provides suggestions on how religion can be cited in support of evolution, and federal Judge John E. Jones declares as a matter of law that there is no conflict between evolution and (true) religion.

CHAPTER IV

THE LIMITED VALUE OF KITZMILLER AS PRECEDENT

Legal scholars debate endlessly the question of how to describe the way in which past legal precedent affects future cases. "Controlling precedent" is an elusive concept. Nonetheless, there are several reasons why Judge Jones' opinion should not be treated as controlling precedent.

A. Cases Deal with the Parties Before Them

It is inherent in the nature of judicial decision making that judges limit the rulings that they make to the parties before them. Only those parties are bound to follow the judgment (including equitable relief such as injunctions) that are imposed by the court. Strictly speaking, even a neighboring school district would not be bound by Judge Jones' opinion that it is illegal to teach intelligent design. Of course, such a school board would be very cognizant of the risk of being sued for teaching intelligent design, but it is only because the original judge's opinion could have *persuasive effect* that it has precedential value anywhere else. However, Judge Jones' opinion does not deserve to have such a persuasive effect because of the multiple errors of fact and interpretation previously documented.

B. An Adverse Judgment against a Party Requires an Opportunity for Them to Be Heard

As mentioned in the Introduction, Discovery Institute, the leading organization supporting research on intelligent design, adamantly opposed the Dover policy and repeatedly urged the school board to withdraw it, beginning long before the district had been sued.[183] Further, three key witnesses affiliated with the Institute withdrew after the law firm representing the school board made unacceptable demands regarding their testimony.[184] Institute leaders also did not see how they could fairly defend a policy with which they disagreed without being tainted by it, a position they made clear in an amicus brief. Additionally, The Foundation for Thought and Ethics (FTE), the publisher of the textbook *Of Pandas and People*, applied to Judge Jones to be granted status as an "intervening party" in the case, but Judge Jones rejected its request.

One might have hoped that the judge rejected FTE's request because he understood that a narrow ruling was called for, both because such a ruling was possible and because the leading proponents of intelligent design were clearly not of one mind with the Dover school board and its counsel. If he had determined to rule narrowly, then obviously his refusal to grant FTE's request would have made a certain amount of sense. But instead he proceeded to rule on the constitutionality of using FTE's textbook in public schools and to criticize the origins and content of the book. Indeed, Judge did not hesitate to write an opinion so broad as to make the most activist of judges envious. He began with an analysis of the "intelligent design movement," which he abbreviated "IDM" because of his frequent references (19) to it and, in a sweeping conclusion, announced that "ID is nothing less than the progeny of creationism."[185] If Judge Jones wanted to address the global issue of whether the intelligent design movement is simply an evolved form of creationism, he should

---

183. Seth L. Cooper, *Statement by Seth L. Cooper Concerning Discovery Institute and the Decision in Kitzmiller v. Dover Area School Board Intelligent Design Case*, Evolution News & Views, at http://www.evolutionnews.org/2005/12/statement_by_seth_l_cooper_con.html (last visited Jan. 30, 2006).

184. *Setting the Record Straight about Discovery Institute's Role in the Dover School District Case*, at http://www.discovery.org/scripts/viewDB/index.php?command=view&id=3003&program=News&callingPage=discoMainPage (last visited Jan. 30, 2006).

185. *Kitzmiller*, 2005 WL 3465563, at *14. The opinion by Judge Jones references Discovery Institute 13 times.

have paid closer attention to whether or not the "movement" had been given a proper opportunity to participate meaningfully in the trial.

C. The Absence of Parties to an Appeal

One mechanism for correcting errors at the trial court level is the availability of the right to appeal on behalf of the party who is the victim of bad judicial reasoning. Of course, many cases are never appealed because the losing party recognizes that the adverse judgment was not a result of legal error, and therefore an appeal would be futile. In this case, by contrast, not only was the "intelligent design movement" never a party to the case, but the board members who represented the nominal defendant (the Dover Area School District) were voted out of office in the November election six weeks before the opinion was issued. The new school board, which has the power to appeal the case, campaigned on a platform that essentially agreed with those who filed the lawsuit. Moreover, they waited to change the policy until *after* the judge issued his opinion—only because they wanted the judge to rule against the former board members' policy and in spite of the legal jeopardy that they created by waiting.[186] As a consequence, there is no party who has any stake in correcting the judge's errors. This is similar to a case in which a trial court makes an erroneous ruling, but before the appellate court can correct the error, the parties settle and the issue becomes moot.

To summarize, the *Kitzmiller* case is legally binding only on the Dover Area School Board. Beyond that, *Kitzmiller* depends for its influence on the persuasiveness of its analysis. As the previous sections have dem-

---

186. If the newly elected board had withdrawn the policy, they could have argued that the case was moot, and could have resisted the claim by the plaintiffs to recover the more than $2 million in attorney fees that were ultimately assessed by Judge Jones. As it was, the district eventually agreed to pay the plaintiffs $1 million dollars to settle their claims. See *School Board Approves Payment to Intelligent Design Legal Team*, PR NEWSWIRE, Feb. 21, 2006, at http://www.prnewswire.com/cgi-bin/stories.pl?ACCT=104&STORY=/www/story/02-21-2006/0004286486&EDATE (last visited Feb. 28, 2006).

onstrated, Judge Jones does not offer a persuasive analysis of the problem and his opinion should be relegated to a historical footnote.

CONCLUSION

The Need to Protect Academic Freedom

Judge Jones' opinion highlights the pressing need to affirm and defend the right of teachers and students to express honest disagreement with the claims of Darwinian evolution. For all of his concern about the illegitimacy of requiring teachers to mention intelligent design or to "denigrate or disparage"[187] evolution, Judge Jones showed no similar interest in the freedom of teachers and students to express opinions that might be critical of Darwinian evolution. As a result, his opinion is likely to be used by defenders of Darwin's theory as a pretext for censoring even completely voluntary expressions of dissenting scientific views by teachers and students.

Teachers seeking to "teach the controversy" over Darwinian evolution in today's climate will likely be met with false warnings that it is unconstitutional to say anything negative about Darwinian evolution. Students who attempt to raise questions about Darwinism, or who try to elicit from the teacher an honest answer about the status of intelligent design theory will trigger administrators' concerns about whether they stand in constitutional jeopardy. A chilling effect on open inquiry is being felt in several states already, including Ohio, South Carolina, and California. Judge Jones' message is clear: give Darwin only praise, or else face the wrath of the judiciary.

Ironically, in the 1980s when the Louisiana Legislature tried to pass an "Academic Freedom Act" to permit teachers to teach "creation sci-

---

187. *Kitzmiller*, 2005 WL 3465563 at *52.

ence," the Supreme Court replied by saying that the announced a purpose of protecting academic freedom was a "sham," because the act "does not give schoolteachers a flexibility that they did not already possess to supplant the present science curriculum with the presentation of theories, besides evolution, about the origin of life."[188] In other words, the Supreme Court thought it was so clear that teachers had the academic freedom to present alternative theories that an act permitting them to do so was superfluous.

After *Kitzmiller*, no one can seriously maintain that academic freedom to study all of the evidence relating to Darwinian evolution is secure. As a consequence, administrative guidelines, even legislative enactments, are needed to provide clearer protection for the rights of students and teachers to critically analyze Darwin's theory in the classroom. Otherwise it is the Supreme Court's own rulings that will be made a "sham."

188. *Edwards*, 482 U.S. at 587.

Whether ID is Science: A Response to the Opinion of the Court in Kitzmiller vs. Dover Area School District

by Michael J. Behe

Introduction

On December 20, 2005 Judge John Jones issued his opinion in the matter of *Kitzmiller*, in which I was the lead witness for the defense. There are many statements of the Court scattered throughout the opinion with which I disagree. However, here I will remark only on section E-4, "Whether ID is Science."

The Court finds that intelligent design (ID) is not science. In its legal analysis, the Court takes what I would call a restricted sociological view of science: "science" is what the consensus of the community of practicing scientists declares it to be. The word "science" belongs to that community and to no one else. Thus, in the Court's reasoning, since prominent science organizations have declared intelligent design to not be science, it is not science. Although at first blush that may seem reasonable, the restricted sociological view of science risks conflating the presumptions and prejudices of the current group of practitioners with the way physical reality must be understood.

On the other hand, like myself most of the public takes a broader view: "science" is an unrestricted search for the truth about nature based on reasoning from physical evidence. By those lights, intelligent design is indeed science. Thus there is a disconnect between the two views of what "science" is. Although the two views rarely conflict at all, the dissonance grows acute when the topic turns to the most fundamental matters, such as the origins of the universe, life, and mind.

Below I proceed sequentially through section E-4. Statements from the opinion are in italics, followed by my comments.

Commentary

1. *ID violates the centuries-old ground rules of science by invoking and permitting supernatural causation.*

It does no such thing. The Court's opinion ignores, both here and elsewhere, the distinction between an implication of a theory and the theory itself. As I testified, when it was first proposed the Big Bang theory struck many scientists as pointing to a supernatural cause. Yet it clearly is a scientific theory, because it is based entirely on physical data and logical inferences. The same is true of intelligent design.

2. *The argument of irreducible complexity, central to ID, employs the same flawed and illogical contrived dualism that doomed creation science in the 1980's.*

The dualism is "contrived" and "illogical" only if one confuses ID with creationism, as the Court does. There are indeed more possible explanations for life than Darwinian evolution and young earth creation, so evidence against one doesn't count as evidence for the other. However, if one simply contrasts intelligent causes with unintelligent causes, as ID does, then those two categories do constitute a mutually exclusive and exhaustive set of possible explanations. Thus evidence against the ability of unintelligent causes to explain a phenomenon does strengthen the case for an intelligent cause.

3. *ID's negative attacks on evolution have been refuted by the scientific community.*

To the extent that the Court has in mind my own biochemical arguments against Darwinism, and to the extent that "refute" is here meant as "shown to be wrong" rather than just "controverted", then I strongly disagree, as I have written in a number of places. If "refute" is just intended to mean "controverted", then that is obvious , trivial, and an injudicious use of language. A "controversial" idea, such as ID, by definition is "controverted."

4. *ID is predicated on supernatural causation, as we previously explained and as various expert testimony revealed. ... (21:96-100 (Behe); P-718 at 696, 700 ("implausible that the designer is a natural entity").*

Again, repeatedly, the Court's opinion ignores the distinction between an implication of a theory and the theory itself. If I think it is implausible that the cause of the Big Bang was natural, as I do, that does not make the Big Bang Theory a religious one, because the theory is based on physical, observable data and logical inferences. The same is true for ID.

5. *ID proponents primarily argue for design through negative arguments against evolution, as illustrated by Professor Behe's argument that "irreducibly complex" systems cannot be produced through Darwinian, or any natural, mechanisms. (5:38-41 (Pennock); 1:39, 2:15, 2:35-37, 3:96 (Miller); 16:72-73 (Padian); 10:148 (Forrest)).*

In my remark here I will focus on the word "cannot." I never said or wrote that Darwinian evolution "cannot" be correct, in the sense of somehow being logically impossible, as the court implies (referencing exclusively to Plaintiffs' expert witnesses). In its use of the word "cannot" the Court echoes the unfair strategy of Darwinists to force skeptics to try to prove a negative, to prove that Darwinism is impossible. However, unlike in mathematics or philosophy, in science one cannot conclusively prove a negative. One can't conclusively prove that Darwinism is false any more than one can conclusively prove that the "ether" doesn't exist. With this unfair strategy, rather than demonstrating empirical plausibility, Darwinists claim that the mere logical possibility that random mutation and natural selection may in some unknown manner account for a system counts in their favor.

In the history of science no successful theory has ever demonstrated that all rival theories are impossible, and neither should intelligent design be held to such an unreasonable, inappropriate standard. Rather, a theory succeeds by explaining the data better than competing ideas.

6. *Professor Behe admitted in "Reply to My Critics" that there was a defect in his view of irreducible complexity because, while it purports to be a challenge to natural selection, it does not actually address "the task facing natural selection." (P-718 at 695). Professor Behe specifically explained that "[t]he current definition puts the focus on removing a part from an already functioning system," but "[t]he difficult task facing Darwinian evolution, however, would not be to remove parts from sophisticated pre-existing systems; it would be to bring together components to make a new system in the first place."*

I "admitted" this "defect" in the definition of irreducible complexity in the context of discussing (in passing, in a long article) a zany hypothetical example that Robert Pennock concocted in his book, *Tower of Babel*. Pennock, a philosopher, wrote that a complex watch could be made by starting with a more complex chronometer (a very precise timepiece used by sailors) and carefully breaking it!—So therefore a watch isn't irreducibly complex! As I testified I have not bothered to address Pennock's point because I regard the example as obviously and totally contrived—it has nothing to do with biologically-relevant questions of evolution. That the words of my article are quoted by the Court without any reference to the context of Pennock's silly example appears invidious and is certainly confused.

7. *Although Professor Behe is adamant in his definition of irreducible complexity when he says a precursor "missing a part is by definition nonfunctional," what he obviously means is that it will not function in the same way the system functions when all the parts are present.*

Yes, it's obvious that's what I meant because that's exactly what I wrote in *Darwin's Black Box*: "An irreducibly complex system cannot be produced directly (that is, by continuously improving the initial function, which continues to work by the same mechanism)." (DBB, p. 39) If it doesn't work the same way when a part is missing, then it can't be produced directly, which is just what I wrote. Nonetheless, I do agree that, for example, a computer missing a critical part can still "function" as, say, a door stop. That hardly constitutes a concession on my part.

4. *Professor Behe excludes, by definition, the possibility that a precursor to the bacterial flagellum functioned not as a rotary motor, but in some other way, for example as a secretory system. (19:88-95, Behe).*

I certainly do not exclude that bald possibility merely by definition. In fact in *Darwin's Black Box*, I specifically considered those kinds of cases. However, I classified those as indirect routes. Indirect routes, I argued, were quite implausible:

> Even if a system is irreducibly complex (and thus cannot have been produced directly), however, one can not definitely rule out the possibility of an indirect, circuitous route. As the complexity of an interacting system increases, though, the likelihood of such an indirect route drops precipitously. (DBB, p. 40)

University of Rochester evolutionary biologist Alan Orr agrees that indirect evolution is unlikely:

We might think that some of the parts of an irreducibly complex system evolved step by step for some other purpose and were then recruited wholesale to a new function. But this is also unlikely. You may as well hope that half your car's transmission will suddenly help out in the airbag department. Such things might happen very, very rarely, but they surely do not offer a general solution to irreducible complexity. (Orr, H. A. "Darwin v. intelligent design (again)." *Boston Review* [Dec/Jan], 28-31. 1996)

There is no strict *logical* barrier to a Darwinian precursor to a bacterial flagellum having functioned as a secretory system and then, by dint of random mutation and natural selection, turning into a rotary device. There is also no absolute *logical* barrier to it having functioned as, say, a structural component of the cell, a light-harvesting machine, a nuclear reactor, a space ship, or, as Kenneth Miller has suggested, a paper weight. But none of these has anything to do with its function as a rotary motor, and so none of them explain that actual ability of the flagellum.

A bare assertion that one kind of complex system (say, a car's transmission) can turn into another kind of complex system (say, a car's airbag) by random mutation and natural selection is not evidence of anything, and does nothing to alleviate the difficulty of irreducible complexity for Darwinism. Children who are taught to uncritically accept such vaporous assertions are being seriously misled.

9. *Notably, the NAS has rejected Professor Behe's claim for irreducible complexity by using the following cogent reasoning:*

    [S]tructures and processes that are claimed to be 'irreducibly' complex typically are not on closer inspection.... The evolution of complex molecular systems can occur in several ways. Natural selection can bring together parts of a system for one function at one time and then, at a later time, recombine those parts with other systems of components to produce a system that has a different function. Genes can be duplicated, altered, and then amplified through natural selection.

The complex biochemical cascade resulting in blood clotting has been explained in this fashion.

Well, that's a fine prose summary of the theory, but there is precious little experimental evidence that random mutation and natural selection can do what the NAS claim they can do. As I testified, in the nineteenth century prominent physicists overwhelmingly believed in the ether, not because of positive evidence for it, but because their theories of light required it. The "ether," however, does not exist. Nor do experiments exist that demonstrate the power of natural selection to make irreducibly complex biochemical systems, either directly or indirectly—proclamations of the National Academy notwithstanding. Again, children who are taught to mistake assertions for experimental demonstrations are being seriously misled.

10. *Professor Behe has applied the concept of irreducible complexity to only a few select systems: (1) the bacterial flagellum; (2) the blood-clotting cascade; and (3) the immune system. Contrary to Professor Behe's assertions with respect to these few biochemical systems among the myriad existing in nature, however, Dr. Miller presented evidence, based upon peer-reviewed studies, that they are not in fact irreducibly complex.*

In this section, despite my protestations the Court simply accepts Miller's adulterated definition of irreducible complexity in which a system is not irreducible if you can use one or more of its parts for another purpose, and disregards careful distinctions I made in *Darwin's Black Box*. The distinctions can be read in my Court testimony. In short, the Court uncritically accepts straw man arguments.

11. *In fact, on cross-examination, Professor Behe was questioned concerning his 1996 claim that science would never find an evolutionary explanation for the immune system. He was presented with fifty eight peer-reviewed publications, nine books, and several immunology textbook chapters about the evolution of the immune system;*

however, he simply insisted that this was still not sufficient evidence of evolution, and that it was not "good enough." (23:19, Behe).

Several points:

1. Although the opinion's phrasing makes it seem to come from my mouth, the remark about the studies being "not good enough" was the cross-examining attorney's, not mine.

2. I was given no chance to read them, and at the time considered the dumping of a stack of papers and books on the witness stand to be just a stunt, simply bad courtroom theater. Yet the Court treats it seriously.

3. The Court here speaks of "evidence for evolution". Throughout the trial I carefully distinguished between the various meanings of the word "evolution," and I made it abundantly clear that I was challenging Darwin's proposed mechanism of random mutation coupled to natural selection. Unfortunately, the Court here, as in many other places in its opinion, ignores the distinction between evolution and Darwinism.

I said in my testimony that the studies may have been fine as far as they went, but that they certainly did not present detailed, rigorous explanations for the evolution of the immune system by random mutation and natural selection—if they had, that knowledge would be reflected in more recent studies that I had had a chance to read (see below).

4. This is the most blatant example of the Court's simply accepting the Plaintiffs' say-so on the state of the science and disregarding the opinions of the defendants' experts. I strongly suspect the Court did not itself read the "fifty eight peer-reviewed publications, nine books, and several immunology textbook chapters about the evolution of the immune system" and determine from its own expertise that they demonstrated Darwinian claims. How can the Court declare that a stack of publications shows anything at all, if the defense expert disputes it and the Court has not itself read and understood them?

In my own direct testimony I went through the papers referenced by Professor Miller in his testimony and showed they didn't even contain the phrase "random mutation"; that is, they assumed Darwinian evolution by random mutation and natural selection was true—they did not even try to demonstrate it. I further showed in particular that several very recent immunology papers cited by Miller were highly speculative, in other words, that there is no current rigorous Darwinian explanation for the immune system. The Court does not mention this testimony.

12. *We find that such evidence demonstrates that the ID argument is dependent upon setting a scientifically unreasonable burden of proof for the theory of evolution.*

Again, as I made abundantly clear at trial, it isn't "evolution" but Darwinism—random mutation and natural selection—that ID challenges. Darwinism makes the large, crucial claim that random processes and natural selection can account for the functional complexity of life. Thus the "burden of proof" for Darwinism necessarily is to support its special claim—not simply to show that common descent looks to be true. How can a demand for Darwinism to convincingly support its express claim be "unreasonable"?

The nineteenth century ether theory of the propagation of light could not be tested simply by showing that light was a wave; it had to test directly for the ether. Darwinism is not tested by studies showing simply that organisms are related; it has to show evidence for the sufficiency of random mutation and natural selection to make complex, functional systems.

13. *As a further example, the test for ID proposed by both Professors Behe and Minnich is to grow the bacterial flagellum in the laboratory; however, no-one inside or outside of the IDM, including those who propose the test, has conducted it. (P-718; 18:125-27 (Behe); 22:102-06, Behe).*

If I conducted such an experiment and no flagellum were evolved, what Darwinist would believe me? What Darwinist would take that as evidence for my claims that Darwinism is wrong and ID is right? As I testified to the Court, Kenneth Miller claimed there was experimental evidence showing that complex biochemical systems could evolve by random mutation and natural selection, and he pointed to the work of Barry Hall on the *lac operon*. I explained in great detail to the Court why Miller was exaggerating, was incorrect, and made claims that Barry Hall himself never did. *However, no Darwinist I am aware of subsequently took Hall's experiments as evidence against Darwinism.* Neither did the Court mention it in its opinion.

The flagellum experiment the Court described above is one that, if successful, would strongly affirm Darwinian claims, and so should have been attempted long ago by one or more of the many, many adherents of Darwinism in the scientific community. That none of them has tried such an experiment, and that similar experiments that were tried on other molecular systems have failed, should count heavily against their theory.

14. *We will now consider the purportedly "positive argument" for design encompassed in the phrase used numerous times by Professors Behe and Minnich throughout their expert testimony, which is the "purposeful arrangement of parts." ... As previously indicated, this argument is merely a restatement of the Reverend William Paley's argument applied at the cell level. Minnich, Behe, and Paley reach the same conclusion, that complex organisms must have been designed using the same reasoning, except that Professors Behe and Minnich refuse to identify the designer, whereas Paley inferred from the presence of design that it was God.*

Again, repeatedly, the Court confuses extra-scientific implications of a scientific theory with the theory itself. William Paley would likely think that the Big Bang was a creative act by God, but that does not make the Big Bang theory unscientific. In fact I myself suspect that the Big Bang may have been a supernatural act, but I would not say that science has determined the universe was begun by God—just that science has determined the universe had a beginning. To reach to a conclusion of God or the supernatural requires philosophical and other arguments beyond science.

Scholarly diligence in making proper distinctions should not be impugned as craftiness. I do not "refuse to identify the designer" as the Court accuses. Starting in *Darwin's Black Box* and continuing up through my testimony at trial, I have repeatedly affirmed that I think the designer is God, and repeatedly pointed out that that personal affirmation goes beyond the scientific evidence, and is not part of my scientific program.

15. *Expert testimony revealed that this inductive argument is not scientific and as admitted by Professor Behe, can never be ruled out. (2:40 (Miller); 22:101 (Behe); 3:99 (Miller)).*

Whether the induction is "scientific" of course depends on the definition of science. The induction is based on reasoning from physical evidence, which in my view does make it scientific. As far as design being "never ruled out," as I explained earlier science never rules anything out as a matter of logic; that is, science can't prove in some absolute sense that something doesn't exist. The task of science is simply to adduce evidence to help support one view or weigh against another.

16. *Indeed, the assertion that design of biological systems can be inferred from the "purposeful arrangement of parts" is based upon an analogy to human design.... Professor Behe testified that the strength of the analogy depends upon the degree of similarity entailed in the two propositions; however, if this is the test, ID completely fails.*

The Court has switched in the space of a paragraph from calling the argument for ID an "inductive argument" to calling it an "analogy." That is a critical confusion. As I testified, the ID argument is an induction, not an analogy. Inductions do not depend on the degree of similarity of examples within the induction. Examples only have to share one or a subset of relevant properties. For example, the induction that, *ceteris pa-*

ribus, black objects become warm in the sunlight holds for a wide range of dissimilar objects. A black automobile and a black rock become warm in the sunlight, even though they have many dissimilarities. The induction holds because they share a similar relevant property, their blackness. The induction that many fragments rushing away from each other indicates a past explosion holds for both firecrackers and the universe (in the Big Bang theory), even though firecrackers and the universe have many, many dissimilarities. Cellular machines and machines in our everyday world share a relevant property—their functional complexity, born of a purposeful arrangement of parts—and so inductive conclusions to design can be drawn on the basis of that shared property. To call an induction into doubt one has to show that dissimilarities make a relevant difference to the property one wishes to explain. Neither the judge nor the Darwinists he uncritically embraces have done that in respect to intelligent design.

17. *Unlike biological systems, human artifacts do not live and reproduce over time. They are non-replicable, they do not undergo genetic recombination, and they are not driven by natural selection. (1:131-33 (Miller); 23:57-59 (Behe)).*

Despite Darwinian claims, none of these factors has ever been shown to account for the molecular machinery of life, so we have no reason to think they affect the induction. (See above.)

18. *For human artifacts, we know the designer's identity, human, and the mechanism of design, as we have experience based upon empirical evidence that humans can make such things, as well as many other attributes including the designer's abilities, needs, and desires. ... Professor Behe's only response to these seemingly insurmountable points of disanalogy was that the inference still works in science fiction movies. (23:73 (Behe)).*

Again, the Court confuses an analogy with an induction. Our knowledge of the nature of the designer is not necessary for a conclusion of design based on induction, any more than knowledge of what caused the Big Bang was necessary before we could inductively conclude that the universe had an explosive beginning. Although the Court appears to disdain science fiction movies, the induction works in science as well. The SETI project (Search for Extraterrestrial Intelligence) is based on our ability to recognize the effects of nonhuman, alien intelligence. It was featured in the science-fiction film *Contact*, for example, based upon a work by Carl Sagan.

19. *This inference to design based upon the appearance of a "purposeful arrangement of parts" is a completely subjective proposition, determined in the eye of each beholder and his/her viewpoint concerning the complexity of a system.*

The court implies that apprehending design is akin to judging if a piece of artwork is attractive—a matter of personal taste. Yet Darwin's theory is widely touted as explaining the strong appearance of design in biology; if such appearance is just a "completely subjective proposition", what is Darwin's theory explaining? The Court neglects to mention that the "completely subjective" appearance of design is—in the view of the adamantly Darwinian evolutionary biologist Richard Dawkins—"overwhelming." I testified to that, to Dawkins' proclamation that "Biology is the study of complicated things that give the appearance of having been designed for a purpose," and to other similar statements. I showed the Court a special issue of the journal *Cell* on "Macromolecular Machines," which contained articles with titles such as "Mechanical Devices of the Spliceosome: Motors, Clocks, Springs, and Things". If strong opponents and proponents of design both agree that biology appears designed, then the appearance should not be denigrated by Judge Jones as subjective.

20. *As Plaintiffs aptly submit to the Court, throughout the entire trial only one piece of evidence generated by Defendants addressed the strength of the ID inference: the argument is less plausible to those for whom God's existence is in question, and is much less plausible for those who deny God's existence. (P-718 at 705).*

As I pointed out in my direct testimony to the Court, the Big Bang theory also was deemed less plausible by some scientists who disliked its supposed extra-scientific implications. I showed the Court an editorial in the prestigious journal *Nature* that carried the title "Down with the Big Bang," and called the Big Bang a "philosophically unacceptable" theory which gave succor to "Creationists." Because real people—including scientists—do not base all of their judgments on strictly scientific reasoning, various scientific theories can be more or less appealing to people based on their supposed extra-scientific implications. It is unfair to suggest ID is unique in that regard.

Conclusion

The Court's reasoning in section E-4 is premised on a cramped view of science; the conflation of intelligent design with creationism; an incapacity to distinguish the implications of a theory from the theory itself; a failure to differentiate evolution from Darwinism; and straw man arguments against ID. The Court has accepted the most tendentious and shopworn excuses for Darwinism with great charity and impatiently dismissed evidence-based arguments for design.

All of that is regrettable, but in the end does not impact the realities of biology, which are not amenable to adjudication. On the day after the judge's opinion, December 21, 2005, as before, the cell is run by amazingly complex, functional machinery that in any other context would immediately be recognized as designed. On December 21, 2005, as before, there are no non-design explanations for the molecular machinery of life, only wishful speculations and Just-So stories.

SELECTED PEER-REVIEWED AND PEER-EDITED PUBLICATIONS SUPPORTING THE THEORY OF INTELLIGENT DESIGN (ANNOTATED)

Scientists and theorists who support the theory of intelligent design have published their work in a variety of appropriate technical venues, including peer-reviewed scientific journals, peer-reviewed scientific books (some published by university presses), peer-edited scientific anthologies, peer-edited scientific conference proceedings and peer-reviewed philosophy of science journals and books. Following is an annotated bibliography of selected technical publications of various kinds that support, develop or apply the theory of intelligent design.

Selected Peer-Reviewed Publications that Directly Support Intelligent Design

1. Stephen C. Meyer, *The Origin of Biological Information and the Higher Taxonomic Categories*, 117(2) PROCEEDINGS OF THE BIOLOGICAL SOCIETY OF WASHINGTON 213–229 (2004).

   Meyer argues that competing materialistic models (Neo-Darwinism, Self-Organization Models, Punctuated Equilibrium and Structuralism) are not sufficient to account for origin of the information neces-sary to build novel animal forms present in the Cambrian Explosion. He proposes intelligent design as an alternative explanation for the origin of biological information and the higher taxa.

2. Lönnig, W.-E. *Dynamic genomes, morphological stasis and the origin of irreducible complexity, in* DYNAMICAL GENETICS 101–119 (2004).

Biology exhibits numerous invariants—aspects of the biological world that do not change over time. These include basic genetic process-es that have persisted unchanged for more than three-and-a-half billion years and molecular mechanisms of animal ontogenesis that have been constant for more than one billion years. Such invariants, however, are difficult to square with dynamic genomes in light of conventional evo-lutionary theory. Indeed, Ernst Mayr regarded this as one of the great unsolved problems of biology. In this paper Dr.Wolf-Ekkehard Lönnig, Senior Scientist in the Department of Molecular Plant Genetics at the Max-Planck-Institute for Plant Breeding Research, employs the design-theoretic concepts of irreducible complexity (as developed by Michael Behe) and specified complexity (as developed by William Dembski) to elucidate these invariants, accounting for them in terms of an intelli-gent design (ID) hypothesis. Lönnig also describes a series of scientific questions that the theory of intelligent design could help elucidate, thus showing the fruitfulness of intelligent design as a guide to further scien-tific research.

3. WILLIAM A. DEMBSKI, THE DESIGN INFERENCE: ELIMINAT-ING CHANCE THROUGH SMALL PROBABILITIES (Cambridge, Cambridge University Press 1998).

This book was published by Cambridge University Press and peer-reviewed as part of a distinguished monograph series, *Cambridge Studies in Probability, Induction, and Decision Theory*. The editorial board of that series includes members of the National Academy of Sciences as well a Nobel laureate. Commenting on the ideas in *The Design Inference*, well-known physicist and science writer Paul Davies remarks: "Dembski's attempt to quantify design, or provide mathematical criteria for design, is extremely useful. I'm concerned that the suspicion of a hidden agenda is going to prevent that sort of work from receiving the recognition it deserves." Quoted in L. Witham, *By Design* (San Francisco: Encounter Books, 2003), p. 149.

4. **DARWIN, DESIGN, AND PUBLIC EDUCATION** (John Angus Campbell & Stephen C. Meyer eds., East Lansing, Michigan State University Press 2003). [Hereafter, "DDPE."]

This is a collection of interdisciplinary essays that addresses the scientific and educational controversy concerning the theory of intelligent design. Accordingly, it was peer-reviewed by a philosopher of science, a rhetorician of science, and a professor in the biological sciences from an Ivy League university. The book contains five scientific articles advancing the case for the theory of intelligent design, the contents of which are summarized below:

a. Stephen C. Meyer, *DNA and the origin of life: Information, specification and explanation, in* DDPE 223–285.

Meyer contends that intelligent design provides a better explanation than competing chemical evolutionary models for the origin of the information present in large bio-macromolecules such as DNA, RNA, and proteins. Meyer shows that the term *information* as applied to DNA connotes not only improbability or complexity but also specificity of function. He then argues that neither chance nor necessity, nor the combination of the two, can explain the origin of information starting from purely physical-chemical antecedents. Instead, he argues that our knowledge of the causal powers of both natural entities and intelligent agency suggests intelligent design as the best explanation for the origin of the information necessary to build a cell in the first place.

b. **Michael J. Behe, *Design in the details: The origin of biomolecular machines, in* DDPE 287–302.**

Behe sets forth a central concept of the contemporary design argument, the notion of "irreducible complexity." Behe argues that the phenomena of his field include systems and mechanisms that display complex, interdependent, and coordinated functions. Such intricacy, Behe argues, defies the causal power of natural selection acting on random variation, the "no end in view" mechanism of neo-Darwinism. Yet he notes that irreducible complexity is a feature of systems that are known to be designed by intelligent agents. He thus concludes that intelligent design provides a better explanation for the presence of irreducible complexity in the molecular machines of the cell.

c. **Paul Nelson & Jonathan Wells, *Homology in biology: Problem for naturalistic science and prospect for intelligent design, in* DDPE 303–322.**

Paul Nelson and Jonathan Wells reexamine the phenomenon of homology, the structural identity of parts in distinct species such as the pentadactyl plan of the human hand, the wing of a bird, and the flipper of a seal, on which Darwin was willing to rest his entire argument. Nelson and Wells contend that natural selection explains some of the facts of homology but leaves important anomalies (including many so-called molecular sequence homologies) unexplained. They argue that intelligent design explains the origin of homology better than the mechanisms cited by advocates of neo-Darwinism.

d. **Stephen C. Meyer et al., *The Cambrian explosion: biology's big bang, in* DDPE 323–402.**

Meyer, Ross, Nelson, and Chien show that the pattern of fossil appearance in the Cambrian period contradicts the predictions or empirical expectations of neo-Darwinian (and punctuationalist) evolutionary theory. They argue that the fossil record displays several features—a hierarchical top-down pattern of appearance, the morphological isolation of disparate body plans, and a discontinuous increase in information content—that are strongly reminiscent of the pattern of evidence found in the history of human technology. Thus, they conclude that intelligent design provides a better, more causally adequate, explanation of the origin of the novel animal forms present in the Cambrian explosion.

e. William A. Dembski, *Reinstating design within science, in* DDPE 403–418.

Dembski argues that advances in the information sciences have provided a theoretical basis for detecting the prior action of an intelligent agent. Starting from the commonsense observation that we make design inferences all the time, Dembski shows that we do so on the basis of clear criteria. He then shows how those criteria, complexity and specification, reliably indicate intelligent causation. He gives a rational reconstruction of a method by which rational agents decide between competing types of explanation, those based on chance, physical-chemical necessity, or intelligent design. Since he asserts we can detect design by reference to objective criteria, Dembski also argues for the scientific legitimacy of inferences to intelligent design.

5. Øyvind Albert Voie, *Biological function and the genetic code are interdependent,* 28(4) Chaos, Solitons and Fractals 1000–1004 (May 2006).

In this article, Norwegian scientist Øyvind Albert Voie examines an implication of Gödel's incompleteness theorem for theories about the origin of life. Gödel's first incompleteness theorem states that certain true statements within a formal system are unprovable from the axioms of the formal system. Voie then argues that the information processing system in the cell constitutes a kind of formal system because it "expresses both function and sign systems." As such, by Gödel's theorem it possesses many properties that are not deducible from the axioms which underlie the formal system, in this case, the laws of nature. He cites Michael Polanyi's seminal essay, *Life's Irreducible Structure*, in support of this claim. Like Polanyi, Voie argues that the information and function of DNA and the cellular replication machinery must originate from a source that transcends physics and chemistry. In particular, since as Voie argues, "chance and necessity cannot explain sign systems, meaning, purpose, and goals," and since "mind possesses other properties that do not have these limitations," it is "therefore very natural that many scientists believe that life is rather a subsystem of some Mind greater than humans."

6. Charles B. Thaxton, et al., THE MYSTERY OF LIFE'S ORIGIN: REASSESSING CURRENT THEORIES (4th ed.; Philosophical Library 1984, Lewis & Stanley 1992).

In this book Thaxton, Bradley and Olsen develop a seminal critique of origin of life studies and develop a case for the theory of intelligent design based upon the information content and "low-configurational entropy" of living systems.

Selected Peer-Reviewed Publications that Support ID Concepts by Citations or Conclusions

1. Michael J. Behe & David W. Snoke, *Simulating Evolution by Gene Duplication of Protein Features That Require Multiple Amino Acid Residues*, 13 PROTEIN SCIENCE 2651–2664 (2004).

In this article, Behe and Snoke show how difficult it is for unguided evolutionary processes to take existing protein structures and add novel proteins whose interface compatibility is such that they could combine functionally with the original proteins. By demonstrating inherent limitations to unguided evolutionary processes, this work gives indirect scientific support to intelligent design and bolsters Behe's case for intelligent design in answer to some of his critics.

2. W.-E. Lönnig & H. Saedler, *Chromosome Rearrangements and Transposable Elements,* 36 ANNUAL REVIEW OF GENETICS 389–410 (2002).

This article examines the role of transposons in the abrupt origin of new species and the possibility of a partly predetermined generation of biodiversity and new species. The authors' approach is non-Darwinian, and they cite favorably the work of design theorists Michael Behe and William Dembski.

3. D.K.Y. Chiu & T.H. Lui, *Integrated Use of Multiple Interdependent Patterns for Biomolecular Sequence Analysis,* 4(3) INTERNATIONAL JOURNAL OF FUZZY SYSTEMS 766–775 (September 2002).

The opening paragraph of this article reads: "Detection of complex specified information is introduced to infer unknown underlying causes for observed patterns. By complex information, it refers to information obtained from observed pattern or patterns that are highly improbable by random chance alone. We evaluate here the complex pattern corresponding to multiple observations of statistical interdependency such that they all deviate significantly from the prior or null hypothesis. Such multiple interdependent patterns when consistently observed can be a powerful indication of common underlying causes. That is, detection of significant multiple interdependent patterns in a consistent way can lead to the discovery of possible new or hidden knowledge."

4. Michael J. Behe, *Self-Organization and Irreducibly Complex Systems: A Reply to Shanks and Joplin,* 67 PHILOSOPHY OF SCIENCE 155–162 (2000).

5. Michael J. Behe, *Reply to my critics: A response to reviews of Darwin's Black Box: The Biochemical Challenge to Evolution*, 16 Biology and Philosophy 685–709 (2001).

In these two articles in peer-reviewed philosophy of science journals, Michael Behe addresses in detail to various criticisms of the arguments and analysis presented in his book *Darwin's Black Box: The Biochemical Challenge to Evolution* (The Free Press, 1996).

Selected Peer-Edited Publications that Support Intelligent Design

1. Scott Minnich & Stephen C. Meyer, *Genetic Analysis of Coordinate Flagellar and Type III Regulatory Circuits, in* Proceedings of the Second International Conference on Design & Nature, Rhodes Greece (M. W. Collins & C. A. Brebbia eds., Boston, WIT Press 2004).

This article underwent conference peer review in order to be included in this peer-edited proceedings. Minnich and Meyer do three important things in this paper. First, they refute a popular objection to Michael Behe's argument for the irreducible complexity of the bacterial flagellum. Second, they suggest that the Type III Secretory System present in some bacteria, rather than being an evolutionary intermediate to the bacterial flagellum, is probably represents a degenerate form of the bacterial flagellum. Finally, they argue explicitly that intelligent design is a better than the Neo-Darwinian mechanism for explaining the origin of the bacterial flagellum.

2. Debating Design: From Darwin To Dna (William A. Dembski & Michael Ruse, eds., Cambridge, United Kingdom, Cambridge University Press 2004) (hereinafter DEBATING DESIGN).

a. William A. Dembski, *The logical underpinnings of intelligent design*, in DEBATING DESIGN 311–330.

In this article, Dembski outlines his method of design detection. In it he proposes a rigorous way of identifying the effects of intelligent causation and distinguishing them from the effects of undirected natural causes and material mechanisms. Dembski shows how the presence of specified complexity or "complex specified information" provides a reliable marker or indicator of prior intelligent activity. He also responds to a common criticism made against his method of design detection, namely that design inferences constitute "an argument from ignorance."

b. Walter L. Bradley, *Information, Entropy, and the Origin of Life*, in DEBATING DESIGN 331–351.

Walter Bradley is a mechanical engineer and polymer scientist. In the mid-1980s he co-authored what supporters of intelligent design consider a seminal critique of origin of life studies in the book *The Mystery of Life's Origins*. Bradley and his co-authors also developed a case for the theory of intelligent design based upon the information content and "low-configurational entropy" of living systems. In this chapter he updates that work. He clarifies the distinction between configurational and thermal entropy, and shows why materialistic theories of chemical evolution have not explained the configurational entropy present in living systems—a feature of living systems that Bradley takes to be strong evidence of intelligent design.

c. Michael J. Behe, *Irreducible complexity: obstacle to Darwinian evolution*, in DEBATING DESIGN 352–370.

In this essay Behe briefly explains the concept of irreducible complexity and reviews why he thinks it poses a severe problem for the Darwinian mechanism of natural selection. In addition, he responds to several criticisms of his argument for intelligent design from irreducible complexity and several misconceptions about how the theory of intelligent design applies to biochemistry. In particular he discusses several putative counterexamples that some scientists have advanced against his claim that irreducibly complex biochemical systems demonstrate intelligent design. Behe turns the table on these counterexamples, arguing that these examples actually underscore the barrier that irreducible complexity poses to Darwinian explanations, and, if anything, show the need for intelligent design.

d. **Stephen C. Meyer, *The Cambrian information explosion: evidence for intelligent design*, in Debating Design 371–391.**

Meyer argues for design on the basis of the Cambrian explosion, the geologically sudden appearance of new animal body plans during the Cambrian period. Meyer notes that this episode in the history of life represents a dramatic and discontinuous increase in the complex specified information of the biological world. He argues that neither the Darwinian mechanism of natural selection acting on random mutations nor alternative self-organizational mechanisms are sufficient to produce such an increase in information in the time allowed by the fossil evidence. Instead, he suggests that such increases in specified complex information are invariably associated with conscious and rational activity—that is, with intelligent design.

Brief of Amici Curiae Biologists and Other Scientists in Support of the Defendants in Kitzmiller v. Dover Area School District

Introduction

Amici curiae are scientists who oppose any attempt to define the nature of science in a way that would limit their ability to follow the evidence wherever it may lead. Since the identification of intelligent causes is a well established scientific practice in fields such as forensic science, archaeology, and exobiology,[1] Amici urge this Court to reject plaintiffs' claim that the application of intelligent design to biology is unscientific. Any ruling that depends upon an outdated or inaccurate definition of science, or which attempts to define the boundaries of science, could hinder scientific progress.

---

1. *See* William A. Dembski, *The Design Inference* (Cambridge University Press, 1998).

Interest of Amici Curiae

Amici are professional scientists who support academic freedom for scientific research into the scientific theory of intelligent design. Some Amici are scientists whose research directly addresses design in biology, physics, or astronomy. Other Amici are scientists whose research does not bear directly upon the intelligent design hypothesis, but feel it is a viable conclusion from the empirical data. Finally, some Amici are skeptics of intelligent design who believe that protecting the freedom to pursue scientific evidence for intelligent design stimulates the advance of scientific knowledge. All Amici agree that courts should decline to rule on the scientific validity of theories which are the subject of vigorous scientific debate.

Selected list of Amici Curiae

Note: Scientists are listed by academic affiliation or doctoral degree. A complete list of all 85 Amici Curiae is attached [at the end of this appendix].

Philip Skell, Member, National Academy of Sciences; Emeritus, Evan Pugh Professor of Chemistry, Pennsylvania State University.

Lyle H. Jensen, Fellow, American Association for the Advancement of Science; Professor (Emeritus), Department of Biological Structures and Department of Biochemistry, University of Washington.

Russell W. Carlson, Professor of Biochemistry and Molecular Biology, Executive Technical Director, Plant and Microbial Carbohydrates, Complex Carbohydrate Research Center, University of Georgia.

Dean H. Kenyon, Emeritus Professor of Biology, San Francisco State University.

Ralph W. Seelke, Professor of Molecular & Cell Biology, University of Wisconsin-Superior.

Gary Maki, Director, Center for Advanced Microelectronics and Biomolecular Research, University of Idaho.

Ronald Larson, George Granger Brown Professor of Chemical Engineering, Chair, Department of Chemical Engineering, University of Michigan.

Gregory J. Brewer, Professor of Neurology, Medical Microbiology, Immunology and Cell Biology, Southern Illinois University School of Medicine

Frederick N. Skiff, Professor, Department of Physics and Astronomy, University of Iowa

Wesley L. Nyborg, Professor of Physics (emeritus), University of Vermont

Michael R. Egnor, Professor and Vice-Chairman, Department of Neurological Surgery, State University of New York at Stony Brook

M. Harold Laughlin, Professor & Chair, Department of Biomedical Sciences, University of Missouri

Bruce D. Evans, Chair, Department of Biology, Huntington University

Wusi Maki, Research Assistant Professor, Department of Microbiology, Molecular Biology, and Biochemistry University of Idaho

Granville Sewell, Professor of Mathematics, University of Texas, El Paso

Christian M. Loch, Ph.D. Biochemistry & Molecular Genetics, University of Virginia

I. Caroline Crocker, Ph.D. Immunopharmacology, University of Southampton

Lisanne D'Andrea-Winslow, Associate Professor of Biology

Northwestern College

Mark E. Fuller, Ph.D. Microbiology, University of California, Davis

Christopher P. Williams, Ph.D. Biochemistry, The Ohio State University

Scott H. Northrup, Professor of Chemistry, Tennessee Tech University

Richard M Anderson, Assistant Professor of Environmental Science and Policy, Duke University

Stephen Meyer, Ph.D. Philosophy of Science, Cambridge University

Jonathan Wells, Ph.D. Molecular & Cell Biology, University of California (Berkeley)

Summary of Argument

Courts are ill-suited to resolve debates over the validity of controversial scientific theories. In particular, the scientific theory of intelligent design should not be stigmatized by the courts as less scientific than competing theories. The advance of scientific knowledge depends on uninhibited, robust investigation seeking the best explanation. Over time, new evidence and new perspectives on existing evidence may require the modification of existing theories or even the abandonment of previously accepted theories that have lost their explanatory power.

The method of identifying intelligent causes is well established in many scientific fields.[2] As a result, Amici assert that the hypothesis of

---

2. *Id.* These areas include archaeology, and the Search for Extra-Terrestrial Intelligence (SETI) Project, which seeks to detect intelligently designed radio signals coming from space.

intelligent design can be an appropriate topic for discussion in a curriculum that addresses biological origins as well as for investigation in the laboratory. Efforts to ban the scientific theory of intelligent design from the classroom, whether by a narrow definition of science or by a discriminatory attack on the personal motives of the scientists conducting scientific research into intelligent design, should be rejected by the Court.

Finally, litigation should not usurp the laboratory or scientific journals as the venue where scientific disputes are resolved. Doubts as to whether a theory adequately explains the evidence should be resolved by scientific debate, not by court rulings. Amici urge the Court to avoid a ruling limiting the nature of science, as it would have far-reaching detrimental effects beyond the schoolhouse doors and into the laboratories and careers of many legitimate scientists.

Argument and Citations of Authority

I. The Nature of Science Is Not a Question To Be Decided by The Courts.

Intelligent design, while admittedly a minority view, is currently being vigorously debated by scientists. For example, Cambridge University Press recently published a volume entitled "Debating Design," in which scientists on both sides of the issue stated their respective cases.[3] Whether or not intelligent design is ultimately widely accepted as the most persuasive explanation for particular scientific phenomena, design theorists have formulated their theory based upon a scientific evaluation of the empirical evidence. The current formulation of intelligent design theory by its proponents, and its application to recent scientific discoveries, is still in its youth compared to many other scientific theories. For that very reason it is premature to conclude that one side has triumphed

---

3. Michael Ruse and William Dembski, eds., *Debating Design* (Cambridge University Press, 2004).

and the other has lost. Simply because one group of scientists favors one interpretation, we must not relegate the other side to a category of "non-scientists" who are ineligible to state their case. Amici strenuously object to appeals to the judiciary to rule on the validity of a scientific theory or to rule on the scope of science in a manner that might exclude certain scientific theories from science. These questions should be decided by scientists, not lawyers.

II. Scientific Progress Depends on an Uninhibited Search for Truth.

a. Dissent within Science Is Healthy.

The scientific enterprise advances when scientists make new discoveries correcting or overturning previously held theories. Scientists in many fields operate under a "paradigm," an overarching theory that provides a framework for interpreting data, performing experiments, and doing further research.[4] Paradigms are typically unquestioned by most scientists and reign over thinking in scientific fields much like established law reigns over a society.

The history of science is replete with examples of novel ideas which were given birth when scientists realized that the empirical data conflicted with reigning paradigms.[5] Scientists who observe data that conflicts with popular scientific paradigms form innovative theories to explain the new data. Scientists propounding these new theories often experience sharp opposition from their peers. It is crucial that advocates of the new scientific theories be granted freedom of inquiry to question reigning scientific ideas if scientific progress is to be possible.

---

4. Thomas Kuhn, *The Structure of Scientific Revolutions* (2nd edit., 1970, University of Chicago Press).

5. *Id.* For example, Einstein's theories of relativity helped explain why Newton's classical laws of motion made inaccurate predictions when dealing with objects moving at very high speeds.

b. Existing Scientific Establishments Are Sometimes Unable to Admit Possibility of Error.

The history of science also reveals that novel scientific theories, even those that prove successful, are often resisted by an "old guard" that defends the long-standing paradigms. Philosophers of science teach that scientists committed to the reigning paradigm engage in "normal science" where scientific dogmas are not questioned.[6] Those practicing "normal science" typically close their ears to dissent:

> No part of the aim of normal science is to call forth new sorts of phenomena; indeed those that will not fit the box are often not seen at all. Nor do scientists normally aim to invent new theories, and they are often intolerant of those invented by others.[7]

Intelligent design fits this historical pattern. It is a relatively young scientific theory, based upon relatively new scientific data, which is currently opposed by many "normal scientists" committed to the Neo-Darwinian paradigm.[8]

This opposition to intelligent design within the scientific establishment is more often based on pride and prejudice than an impartial evaluation of the evidence. A case in point is the resolution opposing intelligent design issued by the board of the American Association for the Advancement of Science (AAAS) in 2002.[9] The AAAS declaration reads like an imperial edict, asserting without any discussion of the evidence that "the ID movement has failed to offer credible scientific evidence to support their claim that ID undermines the current scientifically accepted theory of evolution." Notably, several AAAS board members who voted for the resolution were later unable to cite even one article they had read

6. *Id.*

7. *Id.* at 24.

8. *See* Michael Ruse and William Dembski, *Debating Design* 3-4 (Cambridge University Press, 2004).

9. AAAS News Archives, "AAAS Board Resolution on Intelligent Design Theory," http://www.aaas.org/news/releases/2002/1106id2.shtml (last visited Sept. 23, 2004).

by an intelligent design proponent.[10] In other words, they had voted to condemn intelligent design as unscientific without bothering to investigate it for themselves. The AAAS resolution is little more than a political document that seeks to substitute political consensus for scientific demonstration. When the votes of scientific organizations, acting in a political capacity, are substituted for the give-and-take of public argument and refutation, science loses. To convert such votes into a coercive rule of law would only compound the error. Amici ask the Court not to erode the right of all scientists to pursue scientific inquiry regardless of the views of the current scientific majority.

c. Even Theories that are Eventually Proven Erroneous may Benefit Science by Requiring Reexamination of Long-Held Assumptions.

Whether or not intelligent design is adopted as an explanation for biological origins, science benefits from the competition of alternate hypotheses. Amici see great value to design theory simply because it forces scientists to confront evidence which conflicts with the Neo-Darwinian paradigm, and to finally provide better answers for the origin of highly complex and machine-like biological features.

Even eminent critics of design concede that the possible conclusion of design influences their thinking. The co-discoverer of the structure of DNA, Francis Crick, contended that "[b]iologists must constantly keep in mind that what they see was not designed, but rather evolved."[11] Though himself critical of design, the President of the National Academy of Sciences, Bruce Alberts, has acknowledged that cells resemble human-designed machines:

> The entire cell can be viewed as a factory that contains an elaborate network of interlocking assembly lines, each of which is com-

---

10. John West, "Intelligent design could offer fresh ideas on evolution," *Seattle Post Intelligencer*, Dec. 6, 2002, http://www.seattlepi.nwsource.com/opinion/98810_idrebut06.shtml (last visited Sept, 13, 2005).

11. Francis H. C. Crick, *What Mad Pursuit* 138 (Basic Books 1990).

posed of a set of large protein *machines*. . . . Why do we call the large protein assemblies that underlie cell function protein machines? Precisely because, like machines invented by humans to deal efficiently with the macroscopic world, these protein assemblies contain highly coordinated moving parts.[12]

Evolutionary biologist Ernst Mayr explained that the "core of Darwinism... is the theory of natural selection. This theory is so important for the Darwinian because it permits the explanation of adaptation, the `design' of the natural theologian, by natural means...."[13] Finally, prominent evolutionary biologist and intelligent design critic Richard Dawkins writes that, "[b]iology is the study of complicated things that give the appearance of having been designed for a purpose."[14] Thus evolutionary biologists are sensitive to arguments to design and in fact realize that arguments for design pose challenges to their theories.

Amici reiterate that even incorrect scientific theories advance scientific progress by challenging the scientific community to better explain the natural world. Moreover, dissenting scientific viewpoints should not be suppressed. The freedom of scientists to pursue the scientific evidence to its logical conclusion must be protected so that a better explanation, when it emerges, can be accepted. The Court should oppose any requests to define intelligent design as unscientific or to place it outside of the scope of science.

III. Ad Hominem Attacks on Scientists Should Not Be the Basis for Excluding their Scientific Claims.

As this litigation demonstrates, opponents of intelligent design frequently resort to *ad hominem* attacks, asserting that because some sci-

---

12. Bruce Alberts, "The Cell as a Collection of Protein Machines: Preparing the Next Generation of Molecular Biologists," *Cell*, 92: 291, February 6, 1998 (emphasis in original).

13. Ernst Mayr, *Foreword*, Michael Ruse, *Darwinism Defended* xi-xii (1982).

14. Richard Dawkins, *The Blind Watchmaker* 1 (New York: W. W. Norton & Company. 1986).

entists hold religious views, their scientific work should be dismissed as merely "religious."[15] *Creationism's Trojan Horse*, co-authored by Dr. Barbara Forrest (one of plaintiffs' experts), epitomizes the argument that because many intelligent design theorists are devoutly religious, therefore intelligent design proponents intend to pass off religion as science and are not offering design as a scientific theory.[16]

Forrest's book devotes little space to evaluating the science of intelligent design, but is full of documentation of irrelevant connections (sometimes concrete and sometimes highly tenuous) between intelligent design proponents and religious organizations. Such harping upon the religious affiliations of design proponents and their allegedly deceitful scholarship is bigoted as well as beside the point.

This "*Trojan Horse*" method of critique encourages discrimination against intelligent design proponents by fostering a stereotype among academics that supporters of design are incompetent scientists who use deceitful methods to peddle religion as though it were science.[17] Such a prejudicial tactic would never be permitted if the alleged agenda of the accused group were, say, feminism or gay rights. Indeed, no other group of academics face attacks on their professional careers based primarily on their alleged personal beliefs.[18] Arguments employing such *ad hominem* attacks on the supposed religious beliefs of design theorists should be decisively rejected by this Court.

---

15. *See* Barbara Forrest and Paul Gross, *Creationism's Trojan Horse* (Oxford University Press, 2004).

16. "A movement based on religion does not need the credibility afforded by scientific evidence." *Id.* at 314.

17. *See infra* notes 45–51 and accompanying text for documentation of the discrimination leveled at Dr. Guillermo Gonzalez.

18. *See infra,* notes and 35–56 and accompanying text, for a discussion of the discrimination faced by intelligent design sympathizers.

a. Religious Motivations Are Irrelevant to the Scientific Merits of a Hypothesis.

The motivations and religious views of scientists have nothing to do with the scientific validity of their discoveries. For example, the eminent scientists Isaac Newton and Johannes Kepler were devoutly religious and believed God created a rationally comprehensible universe. Despite their religious motivations, their scientific investigations led to accurate explanations of motion which became the bedrock of physical mechanics. Amici thus assert that motivations for conducting scientific investigations have no bearing upon the empirical validity or scientific nature of the conclusions obtained therein.

Additionally, any religious affiliations or beliefs of intelligent design proponents are protected by their First Amendment rights of freedom of religion and association. Regardless of their associations or motivations, intelligent design theorists do not base their arguments on theological premises:

> The design theorists' critique of Darwinism begins with Darwinism's failure as an empirically adequate scientific theory, not with its supposed incompatibility with some system of religious belief. This point is vital to keep in mind in assessing intelligent design's contribution to the creation–evolution controversy. Critiques of Darwinism by creationists have tended to conflate science and theology, making it unclear whether Darwinism fails strictly as a scientific theory or whether it must be rejected because it is theologically unacceptable. Design theorists refuse to make this a Bible-science controversy. Their critique of Darwinism is not based upon any supposed incompatibility between Christian revelation and Darwinism.[19]

Highly probative of this account is the fact that notable sympathizers of intelligent design are not religious. For example, the famous British

---

19. William A. Dembski, *Intelligent Design: The Bridge Between Science and Theology* 112 (InterVarsity Press, 1999).

atheist, Antony Flew, announced in 2004 that he had been persuaded by the empirical data supporting design. Although Flew continued to espouse no religious commitments after his intellectual shift, he stated "[i]t now seems to me that the findings of more than fifty years of DNA research have provided materials for a new and enormously powerful argument to design."[20] This Court should rule that the motivations and religious beliefs of design proponents are irrelevant to the empirical validity or epistemological nature of design theory.

b. Scientists and Advocates on All Sides of this Issue have Religious (or Anti-Religious) Motivations.

Although Amici emphasize that the religious beliefs and motivations of scientists are irrelevant when evaluating the scientific nature of their arguments, Amici feel compelled to point out that leading opponents of intelligent design are not without their own religious (or anti-religious) motivations.

For example, Eugenie Scott, director of a leading activist organization opposing the teaching of design, the National Center for Science Education ("NCSE"), is a "Notable Signer" of the "Humanist Manifesto III." The Manifesto makes broad theological (or "anti-theological") claims that "[h]umans are... the result of unguided evolutionary change. Humanists recognize nature as self-existing."[21]

Another public opponent of intelligent design is Nobel Laureate Steven Weinberg.[22] Weinberg explains his scientific career is motivated by a desire to disprove religion:

---

20. See http://www.biola.edu/antonyflew/page2.cfm (last visited Sept. 10, 2005).

21. Humanist Manifesto III Public Signers, http://www.americanhumanist.org/3/HMsigners.htm (last visited Sept. 10, 2005); Humanism and its Aspirations, http://www.americanhumanist.org/3/HumandItsAspirations.htm (last visited Sept. 10, 2005).

22. Dr. Weinberg testified in support of teaching only the evidence for evolution before the Texas State Board of Education. See Forrest Wilder, "Academics need to get more involved," Opinion, *The Daily Texan*, Oct. 2, 2003. http://

I personally feel that the teaching of modern science is corrosive of religious belief, and I'm all for that! One of the things that in fact has driven me in my life, is the feeling that this is one of the great social functions of science—to free people from superstition.[23]

Lest there be any doubt about Weinberg's meaning, he expresses his hope that "this progression of priests and ministers and rabbis and ulamas and imams and bonzes and bodhisattvas will come to an end, that we'll see no more of them. I hope that this is something to which science can contribute and if it is, then I think it may be the most important contribution that we can make."[24]

Plaintiff's expert Barbara Forrest is on the Board of Directors of the New Orleans Secular Humanist Association (NOSHA).[25] NOSHA is also an affiliate of the Council for Secular Humanism which it describes as "North America's leading organization for non-religious people."[26] NOSHA's links page boasts "The Secular Web," whose "mission is to defend and promote metaphysical naturalism, the view that our natural world is all that there is, a closed system in no need of an explanation and sufficient unto itself."[27] Most notably, NOSHA is an associate member of the American Humanist Association, which publishes the Humanist Manifesto III.[28] In 1996, this American Humanist Association named

---

www.dailytexanonline.com/media/paper410/news/2003/10/02/Opinion/Academics.Need.To.Get.More.Involved-510574.shtml (last visited Sept. 15, 2005).

23. "Free People from Superstition," http://www.ffrf.org/fttoday/2000/april2000/weinberg.html (last visited Sept. 15, 2005).

24. *Id.*

25. NOSHA Who's Who, http://www.nosha.secularhumanism.net/whoswho.html (last visited Sept. 10, 2005).

26. *Id.*

27. *Id.*

28. *See* http://www.americanhumanist.org/3/HumandItsAspirations.htm (last visited Sept. 10, 2005).

Richard Dawkins as its "Humanist of the Year."[29] To help underscore the anti-religious mindset of these humanist organizations, in his acceptance speech for the award before the American Humanist Association, Dawkins stated that "faith is one of the world's great evils, comparable to the smallpox virus but harder to eradicate."[30]

Even the eminent National Academy of Sciences, which has issued various booklets against teaching intelligent design,[31] has a membership of biologists who (according to surveys) are 95% atheists or agnostics.[32] Amici detail these affiliations not because religious (or anti-religious) beliefs are relevant to a scientific argument, but to demonstrate that the legal rule proposed by the plaintiffs would jeopardize the scientific contributions of many critics of intelligent design just as much as the contributions of some intelligent design proponents. It would also inspire a never-ending succession of irrelevant *ad hominem* attacks. Amici urge the Court to reject such a deeply flawed rule that is so inimical to free inquiry.

IV. Efforts To Discriminate Against Intelligent Design Proponents Have Already Begun.

The concern that acceptance of the plaintiffs' claims could adversely affect the freedom of scientists to pursue the truth is hardly a remote contingency. The Court should be aware that opponents of intelligent design, including some of the witnesses testifying in this case, already have sought to hinder the careers and academic freedom of scientists

29. *See* http://www.thehumanist.org/humanist/articles/dawkins.html (last visited Sept. 10, 2005).

30. *Id.*

31. *See* National Academy of Sciences, *Teaching about Evolution and the Nature of Science and Science and Creationism: A view from the National Academy of Sciences* (National Academy Press, 1998); National Academy of Sciences, *Science and Creationism: A View from the National Academy of Sciences* (2nd edit. National Academy Press 1999).

32. Edward J. Larson and Larry Witham, "Scientists and Religion in America," *Scientific American* 281:88–93, September, 1999.

who are sympathetic towards intelligent design. The following examples demonstrate the potential for the plaintiffs' requested relief to become the basis for further efforts to stifle the intelligent design viewpoint.[33]

Richard Sternberg is a trained evolutionary biologist,[34] and former editor of the peer-reviewed biology journal, *Proceedings of the Biological Society of Washington* (PBSW). As a PBSW editor, in 2004 Dr. Sternberg oversaw the publication of a peer-reviewed technical article which supported the hypothesis of intelligent design.[35] Although the article was reviewed and published using normal procedures,[36] Dr. Sternberg subsequently experienced retaliation by his co-workers and superiors at the Smithsonian, including transfer to a hostile supervisor, removal of his name placard from his door, deprivation of workspace, subjection to work requirements not imposed on others, restriction of specimen access, and loss of his keys.[37] Smithsonian officials also tried to smear Dr. Sternberg's reputation[38] and even investigated his religious and political affiliations in violation of his privacy and First Amendment rights.[39] According to an investigation by the U.S. Office of Special Counsel (OSC), these efforts were aimed at creating "a hostile work environment... with

---

33. For an account of modern-day persecution of scientists, See Gordon Moran, *Silencing Scientists and Scholars in Other Fields: Power, Paradigm Controls, Peer Review, and Scholarly Communication* (Greenwich, Connecticut: Ablex Publishing Corporation 1998).

34. Dr. Sternberg holds Ph.D.'s in molecular evolution and theoretical biology. *See* http://www.rsternberg.net/CV.htm (last visited Sept. 9, 2005).

35. Stephen Meyer, "The Origin of Biological Information and the Higher Taxonomic Categories," *Proceedings of the Biological Society of Washington* 117:213–239, 2004.

36. *See* http://www.rsternberg.net/ (last visited Sept. 9, 2005). *See also* http://www.rsternberg.net/OSC_ltr.htm (last visited Sept. 9, 2005).

37. *Id.*

38. Michael Powell, "Editor Explains Reasons for 'Intelligent Design' Article," *Washington Post*, Aug. 19, 2005, A19, http://www.washingtonpost.com/wp-dyn/content/article/2005/08/18/AR2005081801680_3.html (last visited Sept. 15, 2005).

39. *Id.*

the ultimate goal of forcing [Sternberg]... out of the [Smithsonian]."[40] Furthermore, the OSC found that the pro-evolution NCSE helped devise the strategy to have Dr. Sternberg "investigated and discredited."[41] NCSE executive director Eugenie Scott later indicated to the *Washington Post* that Sternberg was lucky he was not fired outright: "If this was a corporation, and an employee did something that really embarrassed the administration... how long do you think that person would be employed?"[42] Dr. Sternberg was singled out because he permitted an open discussion of a dissenting scientific viewpoint, despite the fact that he is neither a proponent of intelligent design nor a creationist.[43]

Another target of intimidation is Guillermo Gonzalez, an astronomer at Iowa State University (ISU). In a recent book, Dr. Gonzalez postulated that the laws of the universe were intelligently designed to permit the existence of advanced forms of life.[44] Some of Dr. Gonzalez's astronomical work fundamental to his design hypotheses appeared on the cover of *Scientific American*.[45] In retaliation against Dr. Gonzalez's application of design to astronomy, his opponents at ISU circulated a petition signed by over 120 faculty members "denouncing 'intelligent

---

40. *See* http://www.rsternberg.net/ (last visited Sept. 9, 2005). See also http://www.rsternberg.net/OSC_ltr.htm (last visited Sept. 9, 2005).

41. *Id.*

42. Michael Powell, "Editor Explains Reasons for 'Intelligent Design' Article," *Washington Post*, Aug. 19, 2005, A19, http://www.washingtonpost.com/wp-dyn/content/article/2005/08/18/AR2005081801680_3.html (last visited Sept. 15, 2005).

43. See Michael Powell, "Editor Explains Reasons for 'Intelligent Design' Article," *Washington Post*, Aug. 19, 2005, A19, http://www.washingtonpost.com/wp-dyn/content/article/2005/08/18/AR2005081801680.html (last visited Sept. 15, 2005).

44. Guillermo Gonzalez and Jay W. Richards, *The Privileged Planet: How Our Place in the Cosmos is Designed for Discovery* (Regnery Publishing, 2004).

45. Guillermo Gonzalez, Donald Brownlee, Peter D. Ward, "Refuges for Life in a Hostile Universe," *Scientific American*, October, 2001.

design..."[46] The leader of the intimidation campaign—also faculty adviser for the ISU Atheist and Agnostic Society[47]—accused Gonzalez of having a hidden religious agenda. Others similarly "charged him with forcing his scientific evidence into a religious prism, fingering him as an academic fraud."[48] Thus the thesis of "religious and cultural agenda"—the *Trojan Horse* stereotype—has spurred efforts to impede scientific research. Like Sternberg, Gonzalez's attempts to focus on science have been futile: "I don't bring God into science. I've looked out at nature and discovered this pattern, based on empirical evidence."[49] After initiating the campaign of harassment, Gonzalez's chief accuser castigated Gonzalez for declining to appear at a "forum" sponsored by critics determined to denounce intelligent design.[50] Since he is coming up for tenure in the near future, Gonzalez is especially vulnerable to this effort to create a hostile work environment.

Other faculty have experienced similar retribution for their prodesign views. Dr. Caroline Crocker was a biology professor at George Mason University until she mentioned intelligent design in a class and was then banned from teaching both intelligent design *and* evolution.[51] Subsequently, her contract was not renewed. Leading design theorist Dr. William Dembski was banned from teaching at Baylor University and

---

46. Jamie Schuman, "120 Professors at Iowa State U. Sign Statement Criticizing Intelligent-Design Theory," *Chronicle of Higher Education*, Aug. 26, 2005, http://www.chronicle.com/temp/email.php?id=7d6oum55u2gs4xgz0zoqckk x4ulkgoy6 (last visited Sept. 9, 2005).

47. *Id.*

48. Reid Forgrave, "Life: A universal debate," *Des Moines Register*, Aug. 31, 2005, http://www.dmregister.com/apps/pbcs.dll/article?AID=/20050831/ LIFE/%20508310325/1001/LIFE (last visited Sept. 12, 2005).

49. *Id.*

50. Lisa Livermore, "'Intelligent design' faces ISU opposition," *Des Moines Register*, Aug. 26, 2005, http://www.desmoinesregister.com/apps/pbcs.dll/ article?AID=/20050826/NEWS02/508260394/1001 (last visited Sept. 9, 2005).

51. Geoff Brumfiel "Cast out from class," *Nature*, 434:1064, Apr. 28, 2005.

forced into a "five-year sabbatical."[52] This followed after Barbara Forrest wrote letters to dissuade scholars from associating with Dembski's Polanyi Center at Baylor because it was "the most recent offspring of the creationist movement."[53] Finally, Dr. Nancy Bryson was removed as head of the Division of Science and Mathematics at Mississippi University for Women, without explanation, the day after she taught an honors forum entitled "Critical Thinking on Evolution."[54] Such incidents have a chilling effect on the freedom of pro-design scientists to voice their scientific views.[55]

By pursuing tactics reminiscent of the McCarthy era, opponents of design have put the integrity of scientific research in jeopardy. These examples illustrate the need for this Court to reject the narrow definition of science proffered by plaintiffs, and to affirm the law's respect for the normal process of scientific debate to generate answers to scientific controversies.

Conclusion

The plaintiffs have invited this Court to determine the status of intelligent design as science. Because the definition of science and the boundaries of science should be left to scientists to debate, this Court should reject the relief requested by the plaintiffs, and affirm the freedom of scientists to pursue scientific evidence wherever it may lead.

---

52. *Id.*

53. Barbara Forrest, Letter to Simon Blackburn, http://www.designinference.com/documents/2005.05.ID_at_Baylor.htm (last visited Sept. 9, 2005).

54. Transcript of Proceedings before Kansas State Board of Education, http://www.ksde.org/outcomes/schearing05072005am.pdf (last visited Sept. 15, 2005).

55. *Id.* This effort to deny academic freedom to intelligent design proponents is fostered by rhetoric from the leading critics of intelligent design. In *Creationism's Trojan Horse*, for example, Forrest and Gross express a "final hope [ ] that readers will consider seriously the question of what they ought to be doing about" the supposed threat from intelligent design. Barbara Forrest and Paul Gross, *Creationism's Trojan Horse* 315 (Oxford University Press, 2004).

Respectfully submitted,

David K. DeWolf, Esquire, Professor of Law

Gonzaga University School of Law

Counsel of Record

L. Theodore Hoppe, Jr., Esquire

C. Scott Shields, Esquire

Shields and Hoppe, Media, PA 19063

Complete List of Amici Curae

Richard M Anderson, Assistant Professor of Environmental Science and Policy, Duke University

Phillip A. Bishop, Professor of Kinesiology, University of Alabama

John A. Bloom, Professor of Physics, Biola University

William H. Bordeaux, Professor of Chemistry, Huntington University

Gregory J. Brewer, Professor of Neurology, Medical Microbiology, Immunology and Cell Biology, Southern Illinois University School of Medicine

Rudolf Brits, Ph.D. Nuclear Chemistry, University of Stellenbosch, South Africa

Mary A. Brown, DVD (Veterinary Medicine), The Ohio State University

John B. Cannon, Ph.D. Chemistry, Princeton University

Russell W. Carlson, Professor of Biochemistry and Molecular Biology, Executive Technical Director, Plant and Microbial Carbohydrates, Complex Carbohydrate Research Center, University of Georgia

Jarrod W. Carter, Ph.D. Bioengineering, University of Washington

Mark A. Chambers, Ph.D. Virology, University of Cambridge

I. Caroline Crocker, Ph.D. Immunopharmacology, University of Southampton

Lisanne D'Andrea-Winslow, Associate Professor of Biology, Northwestern College

Paul S. Darby, M.D., Georgetown University School of Medicine, Ph.D., Organic Chemistry, University of Georgia

Lawrence DeMejo, Ph.D. Polymer Science and Engineering, University of Massachusetts at Amherst

David A. DeWitt, Ph.D. Neuroscience, Case Western University

Michael R. Egnor, Professor and Vice-Chairman, Department of Neurological Surgery, State University of New York at Stony Brook

Bruce D. Evans, Chair, Department of Biology, Huntington University

Kenneth A. Feucht, Ph.D. Anatomy, University of Illinois in Chicago

Clarence Fouche, Professor of Biology, Virginia Intermont College

Mark E. Fuller, Ph.D. Microbiology, University of California, Davis

Charles M. Garner, Professor of Chemistry, Baylor University

Theodore W. Geier, Ph.D. Forrest Hydrology, University of Minnesota

Dominic M. Halsmer, Ph.D. Mechanical Engineering, UCLA

Jeffrey H. Harwell, Conoco/DuPont Professor of Chemical Engineering, The University of Oklahoma

Christian Heiss, Post-Doctoral Associate, Complex Carbohydrate Research Center, University of Georgia

Dewey H. Hodges, Professor of Aerospace Engineering , Georgia Institute of Technology

Curtis Hrischuk , Ph.D. Computer and Systems Engineering, Carleton University, Ottawa, Ontario, Canada

Tony Jelsma, Associate Professor of Biology, Dordt College

Lyle H. Jensen, Professor (Emeritus), Department of Biological Structures and Department of Biochemistry, University of Washington, Fellow, American Association for the Advancement of Science

Jerry D. Johnson, Ph.D. Pharmacology & Toxicology, Purdue University

David H. Jones, Professor of Biochemistry & Chair of Department of Chemistry, Grove City College

Michael J. Kelleher, Ph.D. Biophysical Chemistry, University of Ibadan

Dean H. Kenyon, Emeritus Professor of Biology, San Francisco State University

Carl Koval, Full Professor, Chemistry & Biochemistry, University of Colorado, Boulder

Ronald Larson, George Granger Brown Professor of Chemical Engineering, Chair, Department of Chemical Engineering, University of Michigan

Joseph M. Lary, Epidemiologist and Research Biologist (retired), Centers for Disease Control and Prevention

M. Harold Laughlin, Professor & Chair, Department of Biomedical Sciences, University of Missouri

Garrick Little, Ph.D. Organic Chemistry, Texas A & M University

Christian M. Loch, Ph.D. Biochemistry & Molecular Genetics, University of Virginia

Gary Maki, Director, Center for Advanced Microelectronics and Biomolecular Research, University of Idaho.

Wusi Maki, Research Assistant Professor, Department of Microbiology, Molecular Biology, and Biochemistry University of Idaho

Graham Dean Marshall, Ph.D. Analytical Chemistry, University of Pratoria, South Africa

L. Whit Marks, Emeritus Professor of Physics , University of Central Oklahoma

Thomas H. Marshall, Adjunct Professor, Food, Agricultural and Biological Engineering, The Ohio State University

David A. McClellan, Assistant Professor of Family & Community Medicine, Texas A&M University Health Science Center

Charles H. McGowen, Assistant Professor of Medicine, Northeastern Ohio Universities College of Medicine

David B. Medved, Ph.D. in Physics, University of Pennsylvania

Stephen Meyer, Ph.D. Philosophy of Science, Cambridge University

Ruth C. Miles, Dean of the School of Arts and Sciences, Professor of Chemistry, Malone College

Dr. John Millam, Ph.D. Theoretical Chemistry, Rice University

Forrest M. Mims, Atmospheric Researcher, Geronimo Creek Observatory

Paul J. Missel, Ph.D. Physics, Massachusetts Institute of Technology

K. David Monson , Ph.D. Analytical Chemistry, Indiana University

Dr. Ed Neeland, Assoc. Prof. of Chemistry, Barber School of Arts and Sciences, University of British Columbia

Benjamin K. Nelson, Research Toxicologist (Retired), National Institute for Occupational Safety and Health Centers for Disease Control and Prevention

Bijan Nemati, Ph.D. High Energy Physics, University of Washington

Arthur J. Nitz, Ph.D. Anatomy and Neurobiology, University of Kentucky, Full Professor of Physical Therapy, University of Kentucky

Scott H. Northrup, Professor of Chemistry, Tennessee Tech University

Wesley L. Nyborg, Professor of Physics (emeritus), University of Vermont

Rafe Payne, Professor of Biology, Department of Biological Sciences, Biola University

Todd Peterson, Ph.D. Plant Physiology, University of Rhode Island

Fazale Rana, Ph.D. Chemistry, Ohio University

John Rickert, Ph.D. Mathematics, Vanderbilt University

Mark A. Robinson, Ph.D. Environmental Science, Lacrosse University

Prof. Paul Roschke, A. P. and Florence Wiley I. Professor, Department of Civil Engineering, Texas A&M University

David W. Rusch, Senior Research Scientist, Laboratory for Atmospheric and Space Physics, University of Colorado

Lennart Saari, Adjunct Professor, Wildlife Biology, University of Helsinki

Fernando D. Saravi, M.D., Ph.D. Medical Sciences School, National University of Cuyo, Professor & Director of the Course of Physiology & Biophysics Department of Morphology & Physiology, Medical Sciences School, National University of Cuyo, Mendoza, Argentina

Ralph W. Seelke, Professor of Molecular & Cell Biology, University of Wisconsin-Superior

Granville Sewell, Professor of Mathematics, University of Texas, El Paso

Theodore J. Siek, Ph. D. Biochemistry, Oregon State University

Arlen W. Siert, Ph.D. Environmental Health, Colorado State University

Philip Skell, Emeritus, Evan Pugh Professor of Chemistry, Pennsylvania State University, Member, National Academy of Sciences

Frederick N. Skiff, Professor, Department of Physics and Astronomy, University of Iowa

Timothy G. Standish, Ph.D. Environmental Biology, George Mason University

Dr. Joseph A. Strada, PhD, Aeronautic Engineering, Naval Postgraduate School

James G. Tarrant, Ph.D., Organic Chemistry, University of Texas at Austin

Mark Toleman, Ph.D. Molecular Microbiology, Bristol University, UK

Jairam Vanamala, Postdoctoral Research Associate. Department of Nutrition and Food Science, Texas A&M University

W. Todd Watson, Assistant Professor of Urban and Community Forestry, Department of Forest Science, Texas A&M University

Jonathan Wells, Ph.D. Molecular & Cell Biology, University of California (Berkeley)

Christopher P. Williams, Ph.D. Biochemistry, The Ohio State University

John W. Worraker, Ph.D. Applied Mathematics, University of Bristol

Henry Zuill, Emeritus Professor of Biology, Union College

Printed in the United States
95486LV00004B/315/A